Math essentials

Best
Practice

Best Practice is the ultimate way to learn valuable practices about life.

Grow your **knowledge** and **confidence** with best practice learning and progression.

Introduction

There are basically just three types of people in the world; those who can count, and those who can't!

Only joking! But learning mathematics is not easy for many people. Some people enjoy facing a complex mathematical challenge whilst other people become confused. Why does this happen?

The answer lies with effective learning of the basics. Some individuals struggle to develop number skills as others succeed in mathematical processes with little apparent stress. Teachers worldwide have achieved good results and poor results from a class of students learning together.

This book takes you on a great journey which started thousands of years ago when mathematics began. As you take this fascinating journey you will learn what mathematics is, and also discover some of the reasons why **mathematics** is relevant and vital to your lifestyle today.

Far from a complicated course, this book is your ideal introduction to mathematics. Page by page, and practice by practice, these best practices will establish a really solid foundation you can build upon.

Your journey will move from numbers and math symbols to the branches of mathematics like arithmetic, percentages, measurement, geometry, and algebra. A bonus in this book is you will learn about our most famous mathematicians of the past 2600 years. This gives an insight as to how math developed over centuries.

These best practices conclude with a chapter about the mathematics of **money**. Handling money, saving your money, and making money all involve your skill and proficiency with basic mathematics.

We all count! 'Math essentials' will build your mathematical knowledge and will improve your numerical skills. Have fun.

Enjoy the journey

Contents
Math essentials

What is Math?

Mathematics is the study of numbers, data, space, and shapes.

The study of MATH uses a sophisticated and abstract system involving mathematical thinking, processes, rules, and symbols.

Mathematics is the science which describes relationships between different numbers and between measurable quantities.

Mathematics is also the academic discipline which studies these numerical relationships.

But mathematics is not solely for the benefit of our academics. We can all use maths to make better decisions and to enjoy a fuller life. Mathematics is for YOU!

10….9….8….7….6….5

4….3….2….1….Liftoff!

Mathematics is fun

Math is all around you! Mathematics is not just about using numbers and peculiar symbols. It affects every part of your life and your lifestyle. Math is fun!

Without math there would be no cars or trains, no bridges, and no tall buildings. We could not travel by plane without some mathematical calculations to work out the aerodynamics of flight. And travel to outer space would simply not be possible without some essential math to decide the right amount of thrust required to escape Earth's gravity.

Math relates to all the things we do in the real world every day. The internet would simply not be possible without some mathematics. Our computers are merely machines for performing math extremely quickly. You will find an aesthetic beauty in studying mathematics. Have lots of fun with all of your best practices.

Only language for all

Mathematics is the only language shared by all the people in our world. This language is the same regardless of culture, gender, age or religion. Calculating the cost of a trolley full of groceries involves the same mathematical process regardless of whether the total is expressed in euro, yen, or dollars. Few people are literate in all of the world's spoken languages. But virtually everyone possesses the ability to be literate in the **shared language** of mathematics.

Your literacy in mathematics is called **numeracy**. And numeracy is the shared language of numbers which connects us all with people in all the continents around our globe. With this language we can explain the mysteries of the universe or measure the distance between Adelaide and Athens. We are also able to build computers, and we can discover cures for diseases. Mathematics is for all of us.

Mathematics is an art

Some people say mathematics is an art. It is a system of **skills** which are used to do such things as calculations, devising proofs, clarifying theories, and solving equations. All mathematicians derive an aesthetic pleasure from their work, which is described as an art form. Mathematics can be a very creative activity.

A mathematician is essentially a maker of **patterns**, much like a poet or a painter. Mathematics is the **art of explanation**. This art is not found in the truth. The art is in the explanation, or in the **argument**.

The ancient art form of mathematics has breathtaking depth and a heartbreaking beauty. It is an art form older than any book, and it is more profound than any poem. Mathematics provides a very **special** way of looking at your world and it also provides you with a growing excitement to be experienced.

Mathematics is an art

Mathematics is a science

Maths is a way of **doing** (an art) and also a way of **learning** (a science). Many people feel mathematics is more of a **science** with its system of **knowledge** relating to numbers, shapes and other abstract entities. The interface between mathematics and its applications in science and engineering has driven much development in mathematics.

Mathematics follows the scientific method using 5 steps: **1- observation**, **2- hypothesis**, **3- experimentation**, **4- testing**, and **5- verification**. Mathematics is an indispensable tool to so many other facets of science. And because nature itself is so mathematical, any science intending to describe nature is completely dependent on mathematics.

This book's author believes mathematics is **both** an art and a science. But is it more art? Or more science? What do you think?

Four Math concepts

Mathematics is the academic study of **four** basic concepts: quantity, structure, space, and change. Most of what we see in life is a combination of two or more of these basic mathematical concepts.

Our English word "mathematics" comes from the Greek word "*máthēma*," which means learning, study, and science. Mathematics ultimately has grown and developed to have the more technical meaning of **mathematical study**.

A mathematician is a person whose main area of study is the field of mathematics. The research required to solve complex mathematical problems may possibly take days or weeks of sustained inquiry.

Mathematicians seek out **patterns**. They formulate conjectures and then resolve the validity of these arguments by showing their mathematical proofs.

Quantity is Arithmetic

Quantity is a basic property of things existing as a range of magnitudes. Being a very fundamental term, **quantity** is used when referring to any type of quantitative properties or attributes of physical things.

In mathematics, quantity is essentially another term for **number**. The definition of quantity is “a specified or indefinite number or amount.” Quantity also looks at the related topics of dimensionality, measurements, proportions, equality, and number systems.

Mass, volume, density, force, height, distance, temperature, distance, and time are all typical examples of quantitative properties. The property of **magnitude** of a mathematical object refers to its size. And the property of quantity in mathematics refers to **arithmetic.** This is the most elementary branch, and it is also the oldest branch, of mathematics.

Structure is Algebra

The structure of something is how all of its parts relate to each other. Structure refers to **how** the thing is assembled.

In mathematic terms, a **structure** (known as a **"set"**) consists of mathematical objects which in some manner attach or relate to the set. A **set** is a collection of distinct objects, and considered to be an object in its own right.

Sets are one of the most fundamental concepts in mathematics. Set theory was developed near the end of the 19th century. Set theory is used as a foundation from which most mathematics is derived.

In the area of abstract algebra, an algebraic structure consists of one or more sets which are closed under one or more operations. Abstract algebra is primarily the study of **algebraic structures** and their properties.

Structure is Algebra

Space is Geometry

Geometry is the mathematics of **shape** and **space**. The study of space originates with geometry. When people think of geometry it is mostly Euclidean geometry which is assumed. Euclidean geometry is essentially flat and non-curved. Basic shapes which can be drawn on a piece of paper are an excellent example of two-dimensional Euclidean geometry.

For over 2000 years the term **Euclidean** was not needed as no other geometry type had been conceived. Euclidean geometry is a mathematical system attributed to brilliant Greek mathematician Euclid. His outstanding book titled ***Elements*** is the first known written systematic discussion of geometry.

Euclidean geometry has two types of measurement scales: 1) **angle,** and 2) **distance**. Angle measurement scale is absolute, but the distance scale is relative.

Change is Analysis

Analysis has its origin in the rigorous formulation of infinitesimal **calculus**. To describe change in the natural sciences, calculus was developed as a powerful tool to investigate and examine this change.

Calculus studies how things **change**. The development of calculus and its relevance to physics and engineering is a big factor in the development of modern science.

The study of real numbers is real **analysis**. Complex **analysis** is a study of complex numbers. Functional **analysis** focuses on spaces of functions. Learning calculus is your gateway to the study of functions and limits, referred to as mathematical analysis. Sounds complex?

But there is good news! Your best practices in this book will focus much more on the other three mathematical concepts of quantity, structure, and space.

Math pupils solve problems

Studying math will result in more than just a profitable career. Students who earn their degree in Mathematics become great problem solvers. They learn how to deal with whatever situation which may arise.

Some **career options** available to you after earning your Mathematics degree include statistician, mathematician, finance analyst, purchasing agent, accountant and a great many other careers. Some students enjoy sharing their great passion for mathematics and choose teaching as their career option.

The very famous astronomer Galileo once declared: "The great book of nature can be read only by those who know the language in which it was written. And that language is mathematics." Successful math students are not just skilled in **solving problems**, but they are also confident with **handling adversity**.

“Math” or “Maths”?

Math and *maths* are equally acceptable abbreviations of mathematics. The main difference is *math* is preferred in the USA and Canada. You will find *maths* is preferred in Britain, Australia, and most other English-speaking areas.

Neither abbreviation is correct or incorrect. One could argue *maths* is better because *mathematics* end in *s*. But one could also argue *math* is better because *mathematics* is just a **mass noun** which happens to end in *s*.

If you were raised in a part of the world where people say *maths*, then *maths* is correct for you. But if you come from a country or region which always said *math*, then *math* is also correct for you.

The Greeks use *mathimatiká*, which means “all the mathematical things.” This got translated to Mathematics. We say **mathematics is fun** (not mathematics are fun). Both *math* and *maths* are OK!

Numbers

Numbers are used all over our world every day - all day! You need a solid understanding of "number sense" more now than ever before.

We now encounter a much greater range of numbers, and in a much wider variety of contexts, than we ever encountered in the past.

Numbers are an essential part of your world. Development of your number sense is gradual, and it begins to develop even before you commence going to school.

In this chapter, let's observe how our numbers originated. Where did our numbers came from? How did how our numbers, and our numeral system, develop over a long and interesting history?

Your fingers are digits

The English word "**digit"** comes from the ancient Latin word "***digita***" – which means fingers. We use the term **digit** as it actually comes from the fact the ten digits of your hands correspond to the ten symbols of the common '**base 10** number system'. This number system refers to decimal digits. And the English word **decimal** is derived from the ancient Latin adjective "***dec***" – which means **ten**.

A digit is a number symbol such as "3" or "8". A digit can be used in numerals with a combination of symbols such as "38". The following digits are the ten digits of **Hindu-Arabic numerals**, in value order:

0 1 2 3 4

5 6 7 8 9

Earliest mathematics

Not much is known of the earliest mathematics on our Earth. But the famous tally stick, the **Ishango Bone** from Africa, has tally marks on it. These marks are suggestive of arithmetic, our most elementary branch of math. The Ishango Bone is believed to be over 20000 years old! This artifact was discovered near the headwaters of the Nile River in an area which was buried in a volcanic eruption.

The oldest written evidence of advanced arithmetic dates back to about 4500 years ago to the Sumerian civilization, where clay tablets display some multiplication and division problems.

It is believed the ancient Egyptians were the first culture to use a "**base 10**" numerical **system**. The Egyptians would **express numbers** by drawing a particular **symbol** as many times as it was needed to describe the **multiple** it represented.

Count in lots of ten

Our earliest number systems grew out of our basic desire to **count**. Counting in groups of **ten** is a practice evidently suggested by the total number of fingers on our two hands. This type of counting has been practiced throughout the world by many cultures over thousands of years.

But the act of writing numbers down is only a comparatively recent innovation. The system we use for actually writing the numbers is a place notation system. This numeral system can simply continue to expand without any limit.

Counting up to 20 can be thought of as two lots of 10. And counting up to **100** is the equivalent of counting **10 lots of 10**. Our decimal system of numeration actually originated in India roughly 1500 years ago. Eventually the decimal system spread all around the world via books, European trade, and colonialism.

From India to Europe

Our ten digits (0, 1, 2, 3, 4, 5, 6, 7, 8 & 9) have descended from the **Hindu-Arabic** numeral system developed in India. Hindu scholars developed the system where a sequence of numerals such as "863" is interpreted as a whole number. These Indian numerals were used by Persian mathematicians in India who passed them on to Arabs further west. And these numerals were eventually transmitted to Europe in the 12th century.

By the 13th century, these Hindu-Arabic numerals became more accepted in European mathematical circles. As trade increased, the numerals began to have common usage by the general public in the 15th century. By the end of the 20th century virtually all non-computerized calculations throughout the world were performed with Arabic numerals. Today they are by far the **most common representation of numbers** in our world.

Numeral: number symbol

All of our counting numbers consist of a series of unique names. The English name for our **first** number is **one**. And the numeral symbol we use to represent the number one is: **1.**

The name of our second number is **two**. Our second numeral is **2**. Adding one unit to two units equals **three** units. And three is represented by the numeral **3**. The following are the names of our first nine numbers and their respective numerals:

One -	**1**	Two -	**2**
Three -	**3**	Four -	**4**
Five -	**5**	Six -	**6**
Seven -	**7**	Eight -	**8**
Nine -	**9**		

Each of the above whole numbers is simply just **one more** than the previous whole number.

Ten & beyond

If you add one more unit to nine units, the name of the next number of units is **ten.** The numerals we use for ten is 10: i.e., the number 1 followed by 0 (zero). Ten is basically a collection of ten ones!

A collection comprising ten lots of ten will add up to the number **one hundred**. Our numeral for the English name of one hundred is: 100.

All of our numbers between ten (10) and twenty (20) have unique names:

11 - Eleven
12 - Twelve
13 - Thirteen
14 - Fourteen
15 - Fifteen
16 - Sixteen
17 - Seventeen
18 - Eighteen
19 - Nineteen

Units of ten

We count our "units of **ten**" as follows:

1 unit of ten	= ten	**10**
2 units of ten	= twenty	**20**
3 units of ten	= thirty	**30**
4 units of ten	= forty	**40**
5 units of ten	= fifty	**50**
6 units of ten	= sixty	**60**
7 units of ten	= seventy	**70**
8 units of ten	= eighty	**80**
9 units of ten	= ninety	**90**

When writing down a numeral for any of the above units of ten; simply follow the number of the unit with a 0 (zero).

Key to system's success

Our system of Hindu-Arabic numerals is very special! The key to the dynamic success of our system of numerals is based on its **positional** nature.

We have ten symbols at our disposal (0, 1, 2, 3, 4, 5, 6, 7, 8, 9), but we are not limited to writing down only these ten different values. Numbers are arranged in columns. When you read columns from **right to left**, the first number represents how many units (**ones**) there are.

The second number from the right-hand side represents how many **tens** there are. And the third number from the right side represents how many **hundreds** there are.

It does matter exactly **which** numeral symbols we select to write down. But it also matters exactly **where** we place all of these numerals in this systematic arrangement of positional numbers.

Additive & multiplicative

Our place value system is both additive and multiplicative. The total value of a particular numeral is found by 1) **multiplying** each place value by its corresponding digit, **and:** 2) then **adding** up all of these resulting products.

Here is an example: Examine the number seven hundred and thirty-two (732). Number 732 has these **place values**:

Place values:	hundred	ten	one
Digits:	7	3	2

Numeral value is: 7 multiplied by 100, plus (3 multiplied by 10), and plus (2 multiplied by 1). So, total value of 732 is: seven hundred, plus thirty, plus two.

Again: (7 x 100) + (3 x 10) + (2 x 1)
= 700 + 30 + 2
= **732**

Prime numbers

A prime number can be **divided only** by **itself and** by **1**. Except for 0 and 1, each of our numbers is **either** a **prime** number or a **composite** number. A composite number is defined as: any number, which is more than one, and not a prime number.

The only even prime number is **2**. All other even numbers can be divided by 2, and these numbers are therefore classed as composite numbers.

No prime number greater than 5 ends in a 5. This is because any number greater than 5 ending in a 5 can obviously be divided by 5.

In order, our first 20 **prime numbers** are:

2 3 5 7 11 13 17
19 23 29 31 37 41 43
47 53 59 61 67 71

Prime numbers

Symbols & Terms

Symbols used in mathematics are universal. Symbols are a concise way of giving lengthy instructions relating to numbers and logic.

We use mathematical symbols to eliminate the need to write long plain language instructions which describe calculations and several other processes. Symbols are not a discovery but an invention. They are a great communication tool.

Let's observe some of the really key symbols you will use daily in all the branches of mathematics.

This chapter also looks at some of the essential mathematical terms you will encounter as you study the vital subject of mathematics. Let's learn the language of mathematics!

Equals (=) Sign

The "**equals sign**" is a mathematical symbol we use to indicate **equality**. This equals sign is placed **between** things stated to be exactly the same:

$$=$$

This = sign is now universally accepted in mathematics as the symbol for equality. The = sign was first recorded by Welsh mathematician Robert Recorde in 1557.

In mathematics, the equals sign is used for **assignment** (an example is: let $x = 6$) and for **testing** or **denoting equality** (an example is: if $x = 6$, then $x \div 2 = 3$).

If two values are equal, we use the equals sign. An example: $4 + 4 = 8$. If two values are **not** equal, we use the "not equals to" sign of: $\neq$. An example of $\neq$ is: $4 + 4 \neq 9$.

Equals (=) sign

Plus (+) & minus (−) signs

The plus sign is written as a **cross** of a horizontal line and a vertical line:

$$+$$

This **plus** sign indicates the **addition** operation of two numbers or expressions. Apart from addition, the plus sign can also be used to indicate a **positive** quantity.

The **minus** sign is used to indicate the mathematical operation of **subtraction**. The inverse of addition is subtraction:

$$-$$

This minus sign can also be used directly in front of a number, when the sign is not a subtraction operator, to mean a **negative** number. For instance, "−4" is minus 4.

JORDAN
23

Multiplication (×) sign

The multiplication sign is mainly used to denote the multiplication of two numbers:

×

This multiplication sign, often named a "times" sign, is used to indicate the preceding number or quantity is to be multiplied by the following number or quantity. Example: $7 \times 5 = 35$, or: $4 \times a = 4a$.

The × symbol was introduced in 1631. But with our computers, the * symbol is often used to denote multiplication because the × symbol does not display on most computer keyboards.

Two terms describe the three numbers involved in a multiplication problem. The **factors** are the two numbers which are multiplied together. The **product** is the end result of this multiplication process.

Division (÷) sign

Division is the arithmetic operation the opposite of multiplication. Division is shown using the **obelus** (division sign):

$$\div$$

The division sign is used to indicate the **preceding number**, or quantity, is to be **divided by** the **following number**, or quantity. Here is an example: $6 \div 3 = 2$; or $2a \div a = 2$.

In division operations, the **dividend** is the number being divided. The **divisor** is the number the dividend will be divided into. In the division example of $6 \div 3 = 2$, the dividend is “6” and the divisor is “3”. The division sign means exactly the same as the **forward slash** sign means. Two examples are: $6 \div 3 = 2$ is the same as: $6/3 = 2$; and $100 \div 10 = 10$ is the same as: $100/10 = 10$.

Fractions (/)

The learning of division usually leads to the concept of fractions being introduced. Dividing two numbers may result in a **remainder**. To complete the operation of division with a remainder, our number system is extended to include fractions.

A fraction is an **expression** representing a part, or several equal parts, of a whole unit. An expression of 2/5 represents just two of the five total parts of the unit have been taken.

All fractions have three components: 1) a **numerator**, 2) a **denominator,** and 3) a **division sign**. In the common example of three-quarters (3/4); the numerator is "3", the denominator is "4", and the division sign, or symbol, is a forward slash "/".

The fraction which represents **one half** is written as: 1/2. And **one third** is: 1/3. **One hundredth** is written as: 1/100.

Decimal point (.)

Another way of representing fractions is by decimal notation. Our decimal system is a numeration system based on powers of 10. Our extremely large numbers are written as a row of digits, with each position in this row corresponding to a particular power of 10.

A decimal point in a row of digits **divides** the number into powers of 10 greater than zero (0), and powers less than 0. Positions of numerals farther to the **right** of the decimal point correspond to increasing **negative** powers. This means the positions extending to the right of a decimal point are increasingly **dividing** by higher powers of 10.

Decimal fractions are often expressed without a denominator. A fraction such as 7/10 (seven tenths) is written as: 0.7; and the fraction of 53/100 (53 hundredths) is denoted by the decimal fraction of: 0.53

Degree (°)

We can measure angles in degrees. There are 360 degrees in one full rotation (one complete circle around).

The **degree** is denoted by a small superscript circle following the number.

(°)

For example 90° means 90 degrees. One degree is 1/360 of a whole circle. This means one full circle is 360°. Note 360 can be divided evenly which makes a lot of your basic geometry easier.

You will learn about degrees and angles in greater detail in the Geometry chapter. Geometry will follow the Measurement chapter in this book. Note "degrees" can also refer to temperature, but in geometry we are referring to angles.

Right angle (∟)

A right angle is an internal angle which is equal to 90°. In geometry, a right angle is an angle which bisects the angle formed by two halves of a straight line. The symbol for a right angle shown below was first used as early as 1698:

∟

As a rotation, a right angle corresponds to a quarter turn. In other words, it is exactly one quarter of a complete circle. The word angle is from the Latin *angulus*, which means: **a corner**.

In geometry, an angle is the figure formed by two endpoints sharing a common endpoint. The magnitude of the angle is the **amount of rotation** which separates the two rays. You will see right angles in many places. One useful place you may see them is in triangles. If a triangle has a right angle, it is called a right triangle.

Pi (π)

The value of π is **approximately** 3.14159. Pi is a mathematical constant whose value is the **ratio** of a circle's circumference to its diameter. Pi is also the same value as the ratio of a circle's area to the square of its radius. In the Greek language, Pi (pronounced "pie") is the first letter of the word for **perimeter**.

$$\pi$$

Many formulae for mathematics, science, and engineering involve π. This of course makes π a very important mathematical constant. A popular approximation of π is 22/7, which is reasonably close, but it is not a completely accurate answer.

π is an **irrational number**. An irrational number is a real number which cannot be expressed as a fraction. And π is probably the most famous of all irrational numbers.

Infinity (∞)

Infinity is not strictly a specific number. The concept of infinity is represented by this unusual and very flowing symbol:

$$\infty$$

This infinity symbol is used to show the concept of an unbounded limit. Infinity refers to a quantity without bound or end.

The symbol for infinity ∞ was chosen by John Wallis in 1655. This ∞ symbol was first found in his *Arithmetica Infinitorum* publication. The infinity symbol ∞ stands for a numerical concept which has given mathematicians numerous problems since the time of the ancient Greeks.

There exists an **infinite** number of **odd** numbers. And of course there are also an **infinite** number of **even** numbers. Our universe itself may be potentially infinite.

Infinity (∞)

Add & Subtract

Arithmetic is the oldest and most elementary branch of mathematics.

Our arithmetic is used for simple day-to-day counting. But it is also used daily for advanced science, industry, and business calculations.

Arithmetic involves the study of quantity, particularly quantity as the result of combining numbers.

The traditional operations of addition and subtraction are two uses of our basic arithmetic.

Let's now look at some of the essential principles which always apply to the operation of addition. Let's also look at the operation of subtraction which is simply the inverse of addition.

Addition combines numbers

Our most basic operation of arithmetic is **addition**. In its simplest form, addition combines just two numbers. These two numbers, referred to as addends, are combined into a single number which is the **sum** of the two numbers.

In the really simple example of $3 + 2 = 5$; the **addends** are 3 and 2, whilst 5 is the **sum** of these two addends.

Adding together more than two numbers can be viewed as repeated addition. The most elementary form of our counting is the repeated addition of the number one.

The basic operation of addition can also be written in a geometric form as follows:

If a and b are the lengths of two separate segments, with these segments placed one after the other, then the **total** length of the two combined segments is the sum: $a + b$.

Practice addition on grid:

+	**1**	**2**	**3**	**4**	**5**	**6**	**7**	**8**	**9**	**10**
1	2	3	4	5	6	7	8	9	10	11
2	3	4	5	6	7	8	9	10	11	12
3	4	5	6	7	8	9	10	11	12	13
4	5	6	7	8	9	10	11	12	13	14
5	6	7	8	9	10	11	12	13	14	15
6	7	8	9	10	11	12	13	14	15	16
7	8	9	10	11	12	13	14	15	16	17
8	9	10	11	12	13	14	15	16	17	18
9	10	11	12	13	14	15	16	17	18	19
10	11	12	13	14	15	16	17	18	19	20

This addition grid shows the **sum** of adding together any two numbers from 1 to 10. Use the above grid to find the sum of two numbers (from top row plus left column) and practice your addition skills.

Choose a number in the **left column** and also a number on the **top row** of the grid. Move across the row of your left number to the box lining up with the top number you chose. The number found in this box is the **sum** of your two numbers!

Adding even / odd numbers

There are some consistent rules which apply when adding even numbers and/or odd numbers. When **adding** two whole numbers together, the following **rules** will always apply:

Even plus **even**	=	an **even** number.
Even plus **odd**	=	an **odd** number.
Odd plus **odd**	=	an **even** number.

Check all of these rules actually do apply by reviewing your addition grid on the previous page. The following rules also apply to the operation of **subtraction**:

Even minus **even**	=	an **even** number.
Even minus **odd**	=	an **odd** number.
Odd minus **odd**	=	an **even** number.

Adding negatives / positives

Numbers can be **positive,** or numbers can be **negative**.

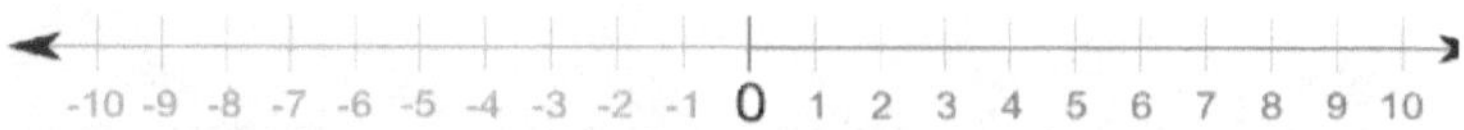

Negative Numbers   Positive Numbers

Adding two positive numbers together is simple addition. And subtracting two positive numbers is simple subtraction.

Subtracting a negative number is the same as adding a positive number! Subtracting a positive number, or adding a negative number, is called **subtraction.**

Two **like** signs together become a **positive** sign, as in these examples: Add a positive; +(+). Or minus a negative; −(−).

Two **unlike** signs become a **negative** sign: Two examples: 1) add a negative; +(−). And 2) minus a positive; −(+)

Law of Commutation

We all understand (1 + 2) gives the same answer as (2 + 1). This equality is true for any two chosen numbers.

So addition is **commutative**. This means the order of the addends being added together does not matter. But this commutation law does not apply to subtraction. If we say "a" and "b" are both numbers, then: $a + b = b + a$, regardless of the respective values of "a" and "b." This is known as the **law of commutation.**

Addition also obeys the **law of association.** With this law, $a + (b + c) = (a + b) + c$. This equality means no matter what order you work out this same sum of numbers, you will get the same answer.

The law of association and the law of commutation are two laws which will both be helpful to you when we explore your algebra section later in this book.

Add with Fibonacci

In mathematics, the Fibonacci numbers are the numbers which are listed in the following sequence:

0, 1, 1, 2, 3, 5, 8, 13, 21, 34, 55, 89, 144, …

By definition, the first two numbers are 0 and 1, and then each subsequent number is the **sum** of the **previous two** numbers.

The Fibonacci sequence of numbers was first introduced to the European society by Leonardo of Pisa, known as Fibonacci, in his 1202 book titled *Liber Abaci*. This sequence had actually been known to Indian mathematicians a long time before.

If you look at the above sequence of Fibonacci numbers, you will see the **difference** between consecutive numbers is the **same** (apart from initial zero) as the preceding number in the above sequence.

Subtract: inverse of Add

Subtraction compares two quantities by taking one number away from another number. There is an **inverse** relationship between addition and subtraction.

If we consider 6 + 4 = 10, then it is true 10 – 6 = 4, and also 10 – 4 = 6. Similar relationships exist for subtraction: If we consider 10 – 2 = 8, then it simply must follow 8 + 2 = 10, and we therefore conclude 2 + 8 = 10.

The reason why the above examples are equal is because we are dealing with an **equation**. An equation is always balanced on either side of the equals (=) sign. If the same thing is done to both sides of the equation, it will remain **balanced**. If we look at 6 + 4 = 10, and subtract the same number from both sides: 6 + 4 – 4 = 10 – 4. On the left side of equals sign, the +4 and the –4 produce 0 (zero). This then leaves us with a balance of 6 = 10 – 4.

“Carry” when adding

When adding two or more numbers which have two digits, align the **place** values in vertical columns. In this addition of two numbers example of:

$$\begin{array}{r} 58 \\ +27 \end{array}$$

Numbers 8 and 7 are shown aligned in the “units” column. And numbers 5 and 2 are aligned in the “tens” column.

First step in adding the above numbers is to add the “**units**” column. 8 + 7 = 15. So “5” is the “units” section of the answer to the above sum. And the “1” in 15 is to be “carried” when adding the “tens” column.

Your next step is to add the numbers shown in the “**tens**” column: 5 + 2 = 7. This actually means 50 + 20 = 70. The last step is to add the “**carried**” 1 to the 7 “**tens**” to now equal 8 in the tens column. So 8 tens plus 5 units = 85.

Subtract: when to "borrow"

When adding numbers, you might "carry" an extra value from one column and move it to the next column. But if subtracting numbers, "borrowing" flips this concept around! You sometimes need to **borrow** a value from a column to subtract.

In subtraction, you borrow when you are subtracting one number which is **greater** than another number is. In the example of (39 – 6) you would obviously not need to use the borrowing concept.

But in the example of (36 – 9) you use borrowing as you cannot easily subtract 9 from 6. When you need to borrow, always look to the column on the **left**. So when subtracting in the "ones" column, and you need to borrow, look to the tens column. When working with tens, borrow from the hundreds. You will only ever **borrow** a "1" and then subtract **one** from the number you borrowed from.

Add or subtract decimals

If you can add and subtract numbers, you can add and subtract decimals! Be sure to place **all** decimal points in a straight line.

When you **add** decimal numbers together: **1**. Place numbers in vertical columns to align decimal points. **2**. Add each column of digits, starting on the right and working left. If the sum of a column is more than 10, "carry" digits to next column on left. **3**. Place the decimal point in the answer **directly** below all the other decimal points in the terms.

To **subtract** decimal numbers: **1**. Place numbers in a vertical column aligning the decimal points. **2**. Subtract each column, starting on the right and working left. If top number is smaller than the bottom number, "borrow" from the top number at the left. **3**. Place decimal point in the answer **directly** below decimal points in the terms.

Pythagoras – born 569 BC

Pythagoras is considered to be one of our first great mathematicians. He is a very key figure in the development of mathematics. Pythagoras of Samos was the head of a society which had an inner circle of followers. This society of scholars was referred to as *mathematikoi*.

Pythagoras was interested in the principles of mathematics, the concept of number, the concept of a triangle, and the abstract idea of a proof. He believed each number had its very own personality – beautiful or ugly, perfect or incomplete. He thought ten was the best number: it contained in itself the first four integers; one, two, three, and four (1 + 2 + 3 + 4 = 10). Pythagoras is today most remembered for his famous **geometry theorem** which bears his name. His theorem relates to a **right-angled triangle**. He is often described as the world's first pure mathematician.

Euclid – born 325 BC

Euclid of Alexandria, Egypt is regarded as the **Father of Geometry**. Euclid is best remembered for his publication of the book *Elements*. These classic works were a compilation of knowledge which became the focus of mathematics for 2000 years!

Euclid's *Elements* is divided into 13 books! Books numbered one to six all deal with plane geometry. And books seven to nine focus on number theory. Book ten deals with the theory of irrational numbers and his remaining books all deal with three-dimensional geometry.
Euclid's classic and enduring publication of ***Elements*** is remarkable for the clarity with which the theorems are stated and proved. It has often been said, with the exception of the Bible, Euclid's *Elements* may be the most translated, published, and studied of all the books ever written in the Western world.

Archimedes – born 287 BC

Archimedes of Syracuse is considered by many to be the greatest mathematician of antiquity. He was able to use infinitesimals in a way similar to modern integral calculus. Archimedes could also provide answers to mathematical problems to an arbitrary degree of accuracy, whilst specifying the limits within which the answer lay. His proof by contradiction is a technique named the **method of exhaustion**.

The proofs of Archimedes are noted not only for their brilliance but also for their unequalled clarity. Archimedes is most renowned for his many theorems of both plane geometry and solid geometry. He discovered formulae for the **volume** and surface **area** of a **sphere**. He may have been first to prove the simple relationship between a circle's circumference and area.

Multiply & Divide

Multiplication and division are also operations of basic arithmetic.

Everything in math is derived from the four basic operations; add, subtract, multiply, and divide.

Multiplication is simply repeated addition! Multiplication is a very fast and convenient way to add a series of numbers.

Division separates a group of objects into a number of equal smaller groups. The results of division can also be obtained by repeated subtraction.

Division is simply multiplication in reverse! Division is a mathematical procedure which will break down a complex problem into easier steps.

Easy numbers to multiply

Easiest number in the world to multiply by is **1**. A number multiplied by 1 equals the number itself. It does **not change.**

The second easiest number to multiply is **10**. Now 10 is a 1 with a zero attached at the end! To multiply a whole number by 10, you multiply by 1 and then simply attach a **zero** at the end. This means the number stays the same except you place a zero at the end. Two examples are: $10 \times 7 = 70$, and $10 \times 34 = 340$.

Another easy number to multiply is 100. You simply attach **two zeros** at the end of the number you started with. An example: $27 \times 100 = 2700$.

To multiply by 20: simply multiply by 10, and then **double** your answer. This works for you because 20 is exactly double the number 10. Some people prefer to double the number **first**, and then multiply by 10.

Multiply is commutative

×	1	2	3	4	5	6	7	8	9
1	1	2	3	4	5	6	7	8	9
2	2	4	6	8	10	12	14	16	18
3	3	6	9	12	15	18	21	24	27
4	4	8	12	16	20	24	28	32	36
5	5	10	15	20	25	30	35	40	45
6	6	12	18	24	30	36	42	48	54
7	7	14	21	28	35	42	49	56	63
8	8	16	24	32	40	48	56	64	72
9	9	18	27	36	45	54	63	72	81

The commutative property of multiplying means it **does not matter** which of your numbers are **first** when you write the problem - the answer is the same: We know 3 x 5 = 5 x 3 (numbers may be switched around and the answer is the same). We know 5 x 9 = 9 x 5. Also 1 x 8 = 8 x 1. And 4 x 10 equals 10 x 4.

Numbers multiplied - factors

A particular number may be produced by multiplying two or more other numbers together. The two or more whole numbers used to multiply together are called **factors** of the final number.

All numbers have a factor of one since one multiplied by any number will naturally equal this number. And all numbers can be divided by themselves to produce the number one. So we normally ignore both the number one and the number itself as useful factors.

If you look at number 15, this number can be divided into just two factors: 3 and 5. But in the example of number 42: 2 × 21 = 42, 3 × 14 = 42, 6 × 7 = 42. So the factors of 42 are: 2, 3, 6, 7, 14 and 21.

All even numbers have a factor of 2, and numbers ending in 5 have a factor of 5. A number ending in 0 has factors of 2 and 5.

Multiplication tricks

To multiply a number by **4**: Double the number, and then double the new number.

To multiply by **5**: Multiply the number by 10, and then divide your answer by 2.

To multiply by **8**: Double the number, double again, and double a third time!

To multiply by **9**: Multiply the number by 10, and then subtract the original number.

To multiply by **12**: Multiply by 10, and then add on twice the original number.

To multiply by **13**: Multiply by 3, and then add on 10 times the original number.

To multiply by **14**: Multiply the number by 7, and then multiply your answer by 2.

To multiply by **15**: Multiply by 10, and then add on 5 times the original number.

To multiply by **16**: Double the number, double the new number again, double it again, and finally double it a fourth time!

To multiply by **19**: Multiply the number by 20, and then simply subtract your original number.

<u>**Multiples of five**</u>

1				**5**					**10**
11				**15**					**20**
21				**25**					**30**
31				**35**					**40**
41				**45**					**50**
51				**55**					**60**
61				**65**					**70**
71				**75**					**80**
81				**85**					**90**
91				**95**					**100**

You will notice each of the bold numbers shown in the above table, which are either ending with a "**5**" or ending with a "**0**", are all **divisible** by 5.

And you will also note if a number ends in a "0", then this number is also divisible by 10. With all of the numbers in bold shown in the above table, you will notice the last digit sequence is: 5, 0, 5, 0, 5, 0, 5, 0, . . . etc.

Squares (^{2}) & square roots (√)

If you first learn all about squares, then you should find square roots reasonably easy. To "square" a number, simply **multiply** the number by **itself**. "Squared" is often written as a little 2 like in this example: $4^2 = 16$. This says, "4 squared equals 16". The little 2 in $\mathbf{4^2}$ is called an **exponent** (or power), and it says the number 4 appears twice in **multiplying**.

Perfect squares are the squares of the whole numbers. Here is a table showing the first ten squares:

1	2	3	4	5	6	7	8	9	10
1	4	9	16	25	36	49	64	81	100

A **square root** goes the other way! A square root of a number is a value which if multiplied by itself, gives the original number. A square root symbol example: $\sqrt{9} = 3$. This says, "square root of 9 equals 3". The square root of 9 is 3; because when 3 is multiplied by itself, you get 9.

Divide is inverse of multiply

Division is the **inverse**, or the **opposite**, of multiplication. If you multiply a number by a particular number and then divide by this same number you multiplied by, the result is you end up with your original starting number.

Because division is the inverse operation of multiplication, the models used to illustrate division are actually the inverse of those models we use to model multiplication.

Division can be understood as separating a set into a number of equal groups. In the example of 12 ÷ 3; this can be viewed as: 12 separated into 3 equal groups:

*	*	*	*

*	*	*	*

*	*	*	*

Numbers divisible by six

To ascertain whether a whole number is evenly divisible by 6 the following rule is used: If a number is **divisible** by **2 and** is **also divisible** by **3**, then this number is also divisible by 6. Let's examine this rule with a couple of different numbers:

First example: number 96 – this number is divisible by 2 as it is an even number (ending in 6). The number 96 is also divisible by 3 as the sum of its two digits is: 9 + 6 = 15, and 15 is divisible by 3. Therefore the number 96 is divisible by 6, as number 96 divides **both** by **2** and by **3**.

Another example is the number 174 – this number is divisible by 2 as it is clearly an even number (ending in 4). The number 174 is also divisible by 3 as the sum of its digits is: 1 + 7 + 4 = 12, and 12 is divisible by 3. So the number 174 is divisible by 6 as it also divides by 2, and the number 174 also divides by 3.

Dividing fractions

There are 3 simple steps to divide fractions: Step 1- Turn the second fraction (the one you want to divide by) **upside-down**. This fraction has now become a **reciprocal**. Step 2- **Multiply** your first fraction listed by this reciprocal.
Step 3- **Simplify** the fraction (if needed).

An example of dividing fractions such as:

$$\frac{1}{2} \div \frac{1}{8}$$

Step 1- Turn the second fraction upside-down. ⅛ turned upside-down becomes 8/1.

Step 2- Multiply the first fraction above by this reciprocal of 8/1:

$$\frac{1}{2} \times \frac{8}{1} = \frac{1 \times 8}{2 \times 1} = \frac{8}{2}$$

Step 3- Simplify the fraction:

$$\frac{8}{2} = 4$$

A fraction tells you to: **multiply** the **top** number, and **divide** the **bottom** number.

Proper & improper fractions

An arithmetic fraction is the division of one number by another.

Fractions are commonly presented in *a/b* form, where *a* and *b* are whole numbers.

In the fraction ⅔, 2 is the numerator, and 3 is the denominator.

If the fraction is less than one (when numerator is less than denominator), it is known as a **proper fraction** (e.g., ⅝, ¾, ½, ¼).

If the fraction is greater than one (numerator larger than denominator) it is known as an **improper fraction** (e.g., 7/6, 13/10, 12/7). In this case the fraction is expressed as a **mixed number** i.e., a combination of a whole number and a proper fraction.

Note that a number cannot be divided by zero.

Rules governing fractions

1) If denominator = numerator, value is 1

2) When multiplying or dividing the numerator and denominator by the same value, the fraction's value doesn't change.

3) Always reduce a fraction to its lowest terms

4) When adding or subtracting fractions, the denominators of all the fractions must be the same.

5) When adding or subtracting fractions with different denominators, fist find the lowest common denominator (LCD) for the fractions

6) To multiply fractions, multiply the numerators and divide this result by the number found by multiplying the denominators

7) To multiply fractions by a whole number, multiply the numerator by the whole number, maintaining the same denominator.

8) To multiply mixed numbers, you must first convert each number to an improper fraction

9) When dividing fractions, invert the dividing fraction and then multiply (as per rule 6)

10) When dividing a whole number by a fraction, invert the fraction and then multiply

Perfect numbers

What defines perfection in a number? For a mathematician, a **perfect number** has a very well-defined mathematical property.

A perfect number is any number whose **divisors** (other than the number itself) add up to equal this same number.

Number 6 is the first perfect number. The divisors of 6 are 1, 2, 3, and 6. If we then add together all the first three numbers (not the number 6), we are left with the sum of: 1 + 2 + 3 = 6. This means 6 fits all criteria to be a **perfect** number.

Our first four perfect numbers are 6, 28, 496, and 8128. All of the perfect numbers which have been found until now end in either a **6** or an **8**. Perfect numbers were deemed to have important numerological properties by many ancient cultures, and they were extensively studied by several notable Greek mathematicians.

Brahmagupta – born 598

Indian mathematician Brahmagupta wrote many notable works on mathematics and astronomy. Nobody gets the unique credit for the invention of our decimal system. But Brahmagupta's famous textbook of 25 chapters, written in 628, *Brahmasphutasiddhanta* (The Opening of the Universe) was very influential. This book is the earliest known text to establish rules for dealing with **zero** as a number in its own right. Brahmagupta made great advances in algebra, arithmetic, numeric analysis, and geometry.

Brahmagupta's mathematics seemed to be influenced by Greek tradition, the achievements of which he generalized and then improved. His mathematics was a mixture of **concrete problems** and **abstract formulas**. All of his writings were delivered to Baghdad, from where they guided the development of the exact sciences in the Islamic world.

Al-Khwarizmi – born 780

Persian mathematician, scientist and author Abu Mohammed ibn-Musa al-Khwarizmi developed the concept of algorithms. He wrote the famous book *Al-Jabr*, which is the origin of our word **algebra**. He delivered significant contributions to trigonometry, astronomy, geography and cartography. Whilst his contributions were the result of original research, he also did much to synthesize the existing knowledge from Greek, Indian, and other sources.

Al-Khwarizmi composed the earliest works on arithmetic and algebra. They were the main source of mathematical knowledge for centuries in both the East and the West. His work on arithmetic introduced Hindu numbers to Europe. After translations into Latin, his works had a profound impact on the advance of mathematics in Europe.

Leonardo Fibonacci – born 1170

Leonardo Pisano Bigollo, also known as Leonardo Fibonacci, or Leonardo of Pisa, was an Italian mathematician considered by many people to be the most talented western mathematician in medieval times. As a young boy, Leonardo journeyed with his merchant father Guglielmo Fibonacci to North Africa (now Algeria). Here he learned about the Hindu-Arabic numeral system. He recognized doing arithmetic with **Hindu-Arabic numerals** was a lot simpler and more efficient than doing arithmetic with Roman numerals.

Fibonacci spent many years studying under the leading Arab mathematicians of his era. At age 32, he published what he had mastered in his *Liber Abaci* (Book of Calculation). This groundbreaking book popularized Arabic numerals in Europe, advocating numeration with all the digits from 0 to 9 and also taught place value.

Percentages

Percentages are an important part of our everyday lives.

Writing something as a percentage is a handy way of writing fractions. But you will notice different percentages can be compared to each other a lot more easily than different fractions can be.

Shops advertise discounts on the products they sell. These discounts are noted as percentages. Financial institutions will quote interest charged to the client on loans as a percentage. The interest paid for money invested will also be quoted as a percentage.

Companies describe their success or failure as a percentage increase or decrease in their profit levels.

Percentage sign

Percentages are used every day to express exactly how large or how small one quantity is, relative to another quantity. Percentages are always denoted by using the number, followed by the percent sign:

%

The fundamental concept of percentages is the percent symbol can be treated as being equivalent to the pure number constant 1/100 (written as a fraction). It is also equal to 0.01 when 1/100 is written as a decimal fraction.

In most writing, we use the percent sign (%) if a number is written as a numeral. Numbers above nine are usually written as numerals. But we write the word "percent" after those numbers which may be written as words. The numbers below 10 are often written as words.

Fraction of 100

In mathematics, a percentage is a way of expressing a number as a **fraction** of **100**. For example, 12% equals 12/100 as a fraction, or 0.12 as a decimal.

The percentage amount is denoted using the **%** sign. When referring to a specific percentage, it is important to specify what the percentage amount is relative to. So, we need to establish the total amount which corresponds to 100% (the whole).

Percent is a mathematical term meaning **one part of a hundred**. "Percent" comes from the Latin words *Per Centum*. The Latin word *Centum* means "one hundred." For example, a "century" is 100 years.

Translations for "per hundred" are *per cento* in Italian, and *pour cent* in French.

A percent is simply a fraction of a whole amount (the whole is always 100%).

Fraction of 100

Fractions as percentages

The fraction one **half** is written as 1/2. The percent equivalent for 1/2 is **50%**.

One **third** is written as 1/3. As a percentage, 1/3 is **33.333....%** whereas 2/3 is 66.666.....% in percentage terms.

One **quarter** is shown as 1/4. Its percent equivalent is **25%**. Three quarters (3/4) have a percent equivalent of 75%.

One **fifth** is written as the fraction 1/5. As a percentage one fifth represents **20%**. This means two fifths, or 2/5, has a percent equivalent of 40%. Three fifths, or 3/5, is 60% and four fifths, or 4/5, has a percent equivalent of 80%.

One **sixth** is noted as 1/6 as a fraction and 16.666.....% as a percentage.

One **eighth** is written as the fraction 1/8, but it is 12.5% in percentage terms.

Formula for percentage

The formula for calculating percentages is relatively simple.

To convert a fraction or a decimal to a percentage, just **multiply by 100**.

To convert a percentage back to a fraction, simply **divide by 100** and then reduce the fraction (if possible).

The simple formula for calculating the percentage is the following:

$$\frac{\text{part}}{\text{whole}} = \frac{\%}{100}$$

One trick to help you calculate percentage is to use the fact *x* percent of *y* is the same as *y* percent of *x*. This rule states 36% of 25 is the same as 25% of 36. Since 25% is 1/4, we note 25% of 36 must be 36/4 or 9. This means 36% of 25 also equals 9.

Percentage of a circle

There are 360 degrees in a full circle. If you construct a circular graph depicting a "pie chart;" you need to establish the percentages and/or angles of the sectors made at the center of your circle.

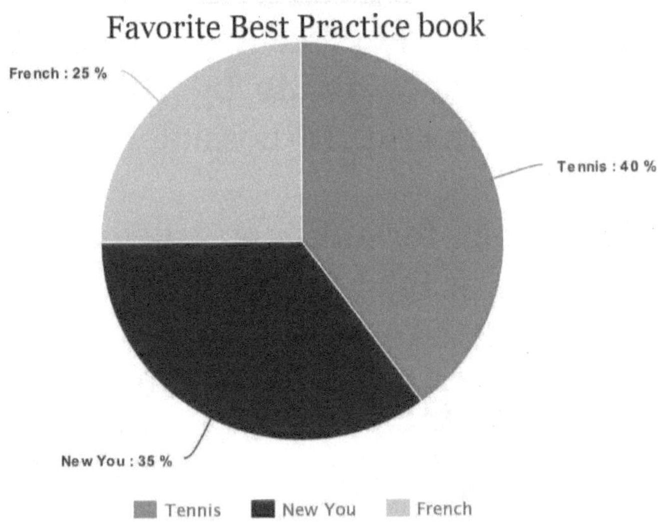

In the above graph, 'Tennis' is shown with 144° (or 40%) of the circle, 'New You' represents 126° (35%), and 'French' is shown with a 90° right angle (25%). The total number of degrees for the three sections of the circle add up to 360°.

Percentage of a number

If you want to know what **percent** A is of B, you simply divide A by B, record the answer's **number** and then just move the decimal place two spaces to the right. This figure is now your **percentage**!

If you need to find 15% of 1400, you will first convert the percentage "15%" to its decimal form: namely, the number "0.15". Then multiply the 0.15 and the 1400. Because 0.15 × 1400 = 210, this says 210 is 15% of 1400.

Look at the question "what percent of 30 is 45? Here we have the original number (30) and the comparative number (45).

The unknown in this problem is the percentage. Since the statement is "45 is some percentage of 30" then the variable stands for the percentage. The equation is: $45 = x$ times 30. This means $45 \div 30 = x = 1.5 = 150\%$. So 45 is 150% of 30.

Percentages as decimals

The % symbol is a quick way to write a fraction with a denominator of 100.

As an example, instead of saying "it rained 17 days out of every 100", we simply say "it rained 17% of the time."

Percentages can be written as decimals by simply moving the decimal point **two places to the left**.

This means decimals can be written as a percentage by moving the decimal point **two places to the right**.

Converting a decimal fraction to a percent is fairly easy. First multiply the decimal fraction by 100. Then simply add a percent sign after the answer! Example: 0.58 as a decimal fraction: firstly, multiply 0.58 by 100 = 58. Now just add a percent sign after the number. So the percentage equivalent of 0.58 is 58%.

Interest calculated with %

If you keep money in a bank, the bank pays you for the use of your money.

The money the bank pays is **interest**. Interest is calculated as a percent of the bank balance. You can find the **amount of interest** by calculating the percentage.

interest rate (% per year) × **principal** = **interest.**

You can also find out the annual **interest rate** by using the following formula:

interest rate × **principal** × **time** = **interest**

Compound interest helps you to grow your principal money faster and faster. The concept of compound interest is interest is added back to the principal sum, so interest is earned on added interest in the next compounding period.

Percentage change

To calculate the percentage increase you need to first work out the difference (amount of increase) between the two numbers you are comparing.

Amount of increase = new number - original number

Then divide the increase by the original number and multiply the answer by 100.

% increase = amount of increase ÷ original number × 100.

If your answer is a negative number then this is a percentage decrease.

To calculate % decrease, first establish the difference (amount of decrease) between the two numbers compared.

% decrease = decrease ÷ original number × 100

Percentage difference

The percentage difference is the **difference** between two values divided by the average of the two values. The difference is shown as a percentage.

Difference means to subtract one value from another. Example: Bob sold 15 tickets, and Dan sold 25 tickets. The difference between 25 and 15 is 10.

Average is the value halfway between:

$$\text{average} = \frac{\text{first value} + \text{second value}}{2}$$

The average of 25 and 15 is: (25 + 15) / 2 = 40/2 = 20.

The find the difference as a **percentage** of the average: Difference is 25 − 15 = 10. Average is (25 + 15)/2 = 20. So 10 as a percentage of 20 is: $\frac{10}{20} \times 100\% =$ **50%**.

Percentage difference

Rene Descartes – born 1596

French mathematician, philosopher and physicist Rene Descartes helped to provide the foundations of our modern calculus. In 1637, Descartes published his innovative and revolutionary work *La Geometrie*. In this outstanding book he combined geometry and algebra to give birth to analytical geometry, which we know today as **Cartesian geometry**.

Descartes is regarded as the first to provide a philosophical framework for the natural sciences as these developed. But his enduring legacy was his development of Cartesian **geometry**. This type of geometry uses a coordinate system on square grid lines using ***x*-axis** and a ***y*-axis**. Each point on the plane has an x-coordinate representing its horizontal position, and a y-coordinate representing vertical position. He also contributed greatly to developing our modern mathematical notation.

Measurement

The greater our precision of measurement, the greater can be our confidence in our observations.

Measurement is finding a number which shows the exact size, or the exact amount, of something.

We measure a lot of things in our civilized world today. We measure quantities of length, mass, time, electric current, luminous intensity, thermodynamic temperature, and amount of substance.

Each of these seven particular quantities have their respective unit of measurement.

Let's now focus on some of the key aspects of measurement and see how you measure up!

International System of Units

The value of a quantity is generally expressed as the product of both a **number** and a **unit**. The unit is a particular example of the quantity referred to, and the number is the ratio of the value of the quantity to the unit.

Because of the importance of a set of well-defined units being universally agreed to, units are chosen so they are readily available to all. Measurements are to be constant throughout time and space.

Our International System of Units is a system of **units of measurement** devised around 7 base units and the convenience of the number 10. The seven quantities of base units are: **length** - meter (m), **mass** - kilogram (kg), **time** - second (s), **amount of substance** - mole (mol) **electric current** - ampere (A), **thermodynamic temperature**- kelvin (K), and **luminous intensity** - candela (cd).

Metric system

Our metric system is a popular system of measurement used by scientists all around the world. Most countries now use only this widely accepted metric system.

This system has prefix modifiers all **multiples of 10**. There are standard ways of talking about big and small numbers:

NAME	NUMBER	PREFIX	SYMBOL
thousand	1000	kilo	k
hundred	100	hecto	h
ten	10	deka	da
unit	**1**		
tenth	0.1	deci	d
hundredth	0.01	centi	c
thousandth	0.001	milli	m

This metric system was first proposed in 1791, and then adopted in France in 1795.

The units and abbreviations used in the metric system shown above are all **interrelated** and are **easy** to remember.

Metric system

Unit of distance - meter

The basic unit of distance in our popular metric system is the **meter** (m). One meter is slightly longer than one yard (an imperial measurement unit of length).

The definition of a meter was originally defined by French scientists to be exactly one ten-millionth of the distance from our equator on Earth to our North Pole!

To measure distances smaller than one meter, the meter can be subdivided into smaller units. One **centimeter** is just one hundredth of a meter, and one **millimeter** measures a tiny one thousandth (1/1000) of a meter.

This of course means there are 10 millimeters in one centimeter: 10mm = 1cm. And so this means there are exactly 100 centimeters in one meter: 100cm = 1m. One kilometer (km) equals 1000 meters: 1km = 1000m.

Area = length × width

The square meter is the basic unit of area in our metric system. **Area** is calculated by multiplying the length by the width. So a square which has a length of one meter on all four sides will measure exactly one square meter in the total area.

This basic unit is **meters** × **meters**, which is written as **m²** (square meters). The area of other shapes can also equal one square meter – an example is a rectangle two meters long and ½ a meter wide. You would calculate this rectangle's area as follows: $2m \times \frac{1}{2}m = 1m^2$ (square meter). The area of a figure is the amount of space contained within this figure.

A hectare is an area equal to a square 100 meters long on all four sides. This means a hectare has a total area measuring $100m \times 100m = 10000m^2$ (square meters). Hectares are commonly used to measure the land area of property.

Area = length × width

Volume = side × side × side

The amount of space occupied by an object is known as its **volume**. This adds a third dimension to all of the objects in the world around us. **Cubic units** are used to measure the total volume of any three-dimensional object.

The basic unit of volume is a cube which measures one meter on all of its sides. The unit is written m^3 (**cubic meters**). A cube exactly one meter long on each of its sides measures exactly one cubic meter.

Liquids are commonly measured by their volume. The two most commonly used measurements of a liquid quantity are liters and milliliters. 1 liter = 1000 milliliters. Liters are often written as **L** (for short), so "2L" means "two liters". The abbreviation for milliliters is written as **ml**. In the abbreviated form, we write: 1L = 1000ml. One cubic meter, m^3 as noted above, is equal to 1000 liters.

Volume

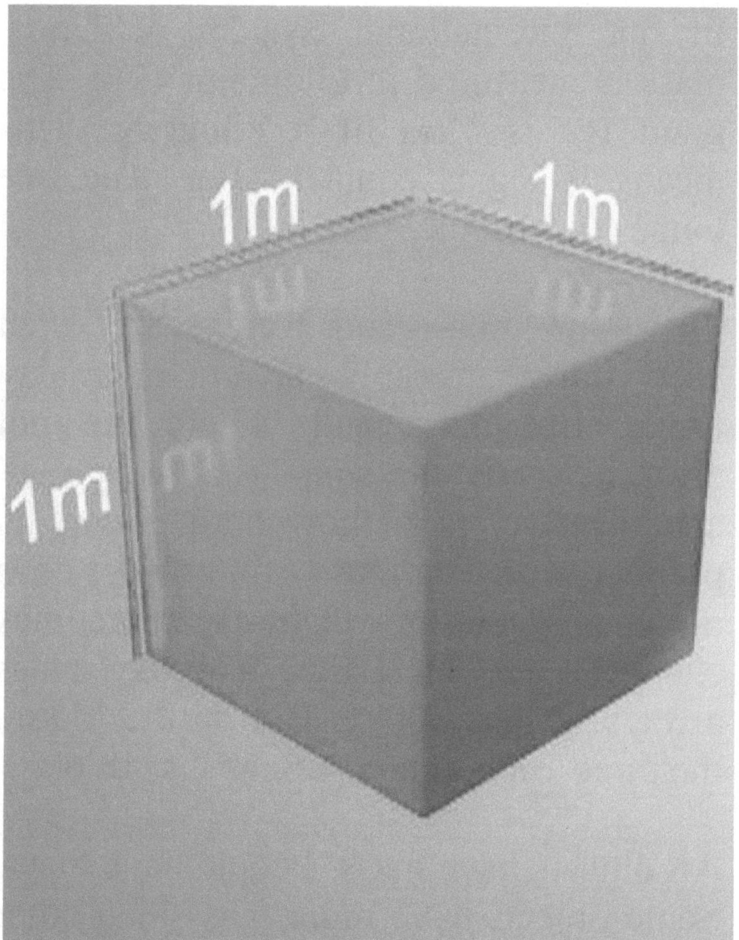

One meter × one meter × one meter = one cubic meter

Image credit: Visual Learning Company (Producer) Cubic meter

Measure mass in kilograms

In our International System of Units, mass is measured in kilograms (kg). The gram (g) is $\frac{1}{1000}$ of a kilogram: This means 1000g is equal to 1kg. And one tonne (t) = 1000kg.

The terms **mass** and **weight** are often used interchangeably in our everyday usage. But mass and weight are not always exactly the same. The weight of an object refers to how hard gravity is pulling on it. The mass of an object does not change, but its weight can sometimes change. Here on Earth, a 100kg mass always weighs 100kg. But on the Moon, the same 100kg mass weighs just 16.6kg!

All objects have mass. People have mass. Some objects have more mass than other objects have. Mass is a measure of how much matter is in an object, or in a given amount of liquid or gas. Matter is defined as anything which has mass and volume.

Scientific standard of time

Measurement of time is the basis of every science, as a scientist can only study what changes with time. The ability to measure time makes our **way of life** possible. And our time is the most accurately measured quantity of any of the physical quantities. Our accuracy in specifying the exact quantity of time is imperative for civil, industrial, and scientific purposes.

Time is one of the seven fundamental physical quantities in our International System of Units (abbreviated as SI). The scientific standard of time is the **second**. The official SI definition of a second is the duration of 9,192,631,770 cycles of the cesium atom's resonant frequency!

From the second, larger units such as the **minute** (60 seconds), **hour** (60 minutes) and **day** (24 hours) are defined. There are no fixed ratios between seconds and years or months, as months and years will vary.

World has 24 time zones

Earth has **360** degrees of longitude around the globe. We have a 24-hour day, and we have **24 time zones** around our planet. If you divide 360° by 24, the answer is 15°. So the width of each individual time zone is an average of just 15°. These individual time zone boundaries are not all perfectly straight lines as some have been adjusted for the convenience of local populations.

In 1884, the line of longitude (meridian) passing through Greenwich in England was adopted as the **prime meridian** for Earth's longitude timekeeping. Located at 180° away from Greenwich in the Pacific Ocean is our **International Date Line**. At this point, the **date changes** from one side of the boundary (eastern side) of this date line to the other side (the western side). Our entire world is experiencing the same date only at the instant of noon in Greenwich, England, when it is exactly midnight at our International Date Line.

Latitude & Longitude

Our planet Earth is very nearly a perfect sphere. On a globe of the Earth, there are two sets of lines which form a special grid used to locate any point on the Earth's surface. These lines are "circles" - great circles and small circles.

Great circles are circles on the surface of a sphere, with greatest possible diameter.

Small circles are circles on the surface of a sphere, with less than the greatest possible diameter.

The **Equator** is a great circle dividing the Earth into northern and southern hemispheres. **Parallels of latitude** are small circles parallel to the equator.

Meridians of longitude are half great circles joining the North and South Poles. Latitude and longitude are measured in degrees and minutes.

Convert Fahrenheit to Celsius

Celsius and Fahrenheit are two different **temperature** scales. A typical use of these temperature scales is to describe what our maximum temperature is expected to be each day. American weather forecasters use the Fahrenheit scale, but the Celsius scale is used in Australia. If you watch overseas news or travel abroad, you will need a formula to convert from one scale to another scale.

When you go **down 5°** in the **Celsius** scale, you go **down 9°** in the **Fahrenheit** scale. If you freeze water, it measures 0° in Celsius, but 32° in Fahrenheit. If you boil water, it measures 100° in Celsius, but it is 212° in Fahrenheit. Mathematical formulas exist for all sorts of different things. The three steps in your **formula** for converting **°F to °C** are 1) **deduct 32**, 2) **multiply** by **5**, and 3) **divide** by **9**. If converting **°C to °F**: 1) **Multiply** by **9**, 2) **divide** by **5**, and 3) **add 32**.

Convert Fahrenheit to Celsius

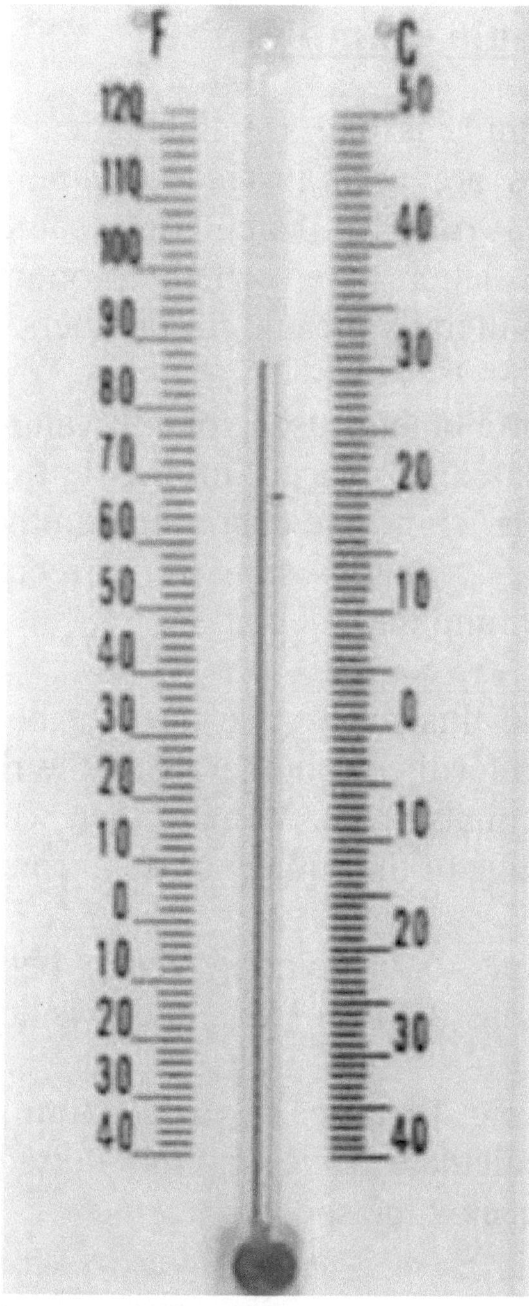
°F
°C
120
110
100
90
80
70
60
50
40
30
20
10
0
10
20
30
40
50
40
30
20
10
0
10
20
30
40

Formula examples

A formula is like a mathematical recipe. Letters are used instead of numbers to give a rule for calculating something. These letters are called 'pronumerals' which literally means 'for numbers.'

The pronumeral used for the value to be calculated is placed to the left of the "equals" sign. The expression on the right of the "equals" sign is a recipe for performing the calculation.

To use this recipe, we need to be given values for the pronumerals on the right of the "equals" sign. We can then work out the value of the subject of the formula.

Example for calculating mass: **M = V x D**
M is mass, V is volume, and D is density.

Example: **Distance = speed x time**
To calculate time, simply divide the distance by the speed.

Convert large to small

The two golden rules of conversion are:

(1) To convert **larger** units **to smaller** units you need to **multiply** (×) by the factor. (2) When converting **smaller** units **into larger** units you simply **divide** (÷).

Metric units of measurement are obtained by multiplying and dividing by powers of 10. Units of mass used are kilogram (kg), gram (g) and milligram (mg). These units relate as follows: 1000mg = 1g. And 1000g = 1kg. To convert 2.45 grams (larger unit) to milligrams (smaller unit), you multiply 2.45 by 1000. This means 2.45g is equal to 2450 mgs.

If a drug dose needed is in milligrams but the stock is in grams, **convert** the grams to milligrams. Calculation is easier if you convert the larger unit (grams), so it is the same as the smaller unit (milligrams).

Sir Isaac Newton – born 1643

Isaac Newton is mostly remembered for his **Three Laws of Motion** (inertia, force, and reciprocal action) and his Law of Universal Gravitation. At the age of 22, Isaac Newton commenced his many revolutionary advances in mathematics, optics, physics, astronomy, dynamics, thermodynamics and acoustics.

Newton developed a type of mathematics now referred to as **calculus**. He developed the foundations for differential and integral calculus. This powerful method carried modern mathematics higher than the level of Greek geometry. Calculus is the mathematical **foundation** on which all of **physics** is based. In 1687 Newton published *Principia*, which is often regarded as the greatest scientific book ever written. Sir Isaac Newton's detailed works on both physics and mathematics completed the scientific revolution of the 17th century.

Geometry

Geometry studies the size, shape, and position of 2-dimensional shapes and 3-dimensional figures.

In geometry you will explore spatial sense and geometric reasoning. Geometry is everywhere.

You will find geometry in art, architecture, engineering, sports, land surveys, astronomy, nature, cars, fashion, and much more.

Geometry is linked to other topics in mathematics. There is a link to measurement used in many jobs such as architects, engineers, and physicists to name just a few.

Your spatial reasoning and problem-solving skills will develop as you learn more about geometry.

Points and Lines

A **point** is one of the most basic concepts in geometry. Points are represented by dots and are named by a capital letter. A point indicates a certain location, and this point can be anywhere in the universe.

Points are dimensionless with no width, length or height. Many physical objects suggest the idea of a point. Well known examples of a point are the corner of a cube, or a dot on a sheet of paper.

We specify a line with two points. A line from one point to the other is known as a **line segment**. A line extends indefinitely in one dimension. Its length is infinite as its ends can be extended forever in both directions. A line has no width or height. You may specify a line by specifying any two points within the line. For any two points, **only one line** passes through **both points**. But an unlimited number of lines may pass through any one single point.

Intersections and planes

The point where two line segments meet is known as an **intersection**. When lines meet they share a **common point** called the point of intersection. Two lines which never meet are said to be parallel.
Here is an example of some intersections:

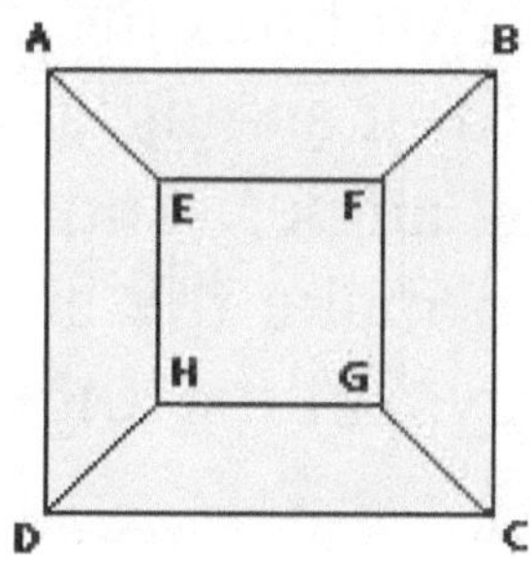

Any line segment which crosses or meets line segment AD is said to **intersect** AD. In the above figure, the line segments of AB, DC, DH, and AE all intersect line segment AD.

A **plane** is defined as an infinite set of points forming a connected **flat** surface extending in all directions. A plane has an infinite length, an infinite width, and has zero height (thickness).

Measure angles in degrees

If two straight lines meet at a point, then an **angle** between these lines is formed. The point of the intersection of these two lines is called the **vertex**. An angle is shown at this intersection, or vertex.

Let's assume two lines meet, or intersect, at point A to form an angle. Point A is the vertex of this angle. And the two lines which meet to make the angle are called the arms, or rays, of the angle.

The **amount** of turn, or rotation, from one arm of this angle to the other arm of this angle is said to be the **size** of this angle.

Angles are all measured in **degrees**. The exact number of degrees explains exactly how wide open this angle is. Degrees are depicted by the ° symbol. There are 360° in a complete circle. Exactly one quarter turn of a full circle measures 90°, which is the measurement of a **right angle**.

Names of angles

Different angles have been given particular names depending on the **size** of the angle:

> An **acute** angle is between 0° and 90°
>
> A **right** angle measures exactly 90°
>
> An **obtuse** angle is between 90° and 180°
>
> A **straight** angle measures exactly 180°
>
> A **reflex** angle is between 180° and 360°
>
> A **full** angle measures 360°

Two angles are said to be **supplementary** angles if these two angles add up to 180°.

Two angles are referred to as being **complementary** if their angles add up to exactly 90°. Note that both these angles would be defined as **acute** (less than 90°).

Squares and rectangles

The two simplest geometrical shapes for you to consider are squares and rectangles. Both a square and a rectangle are part of the class of shapes which are known as **quadrilaterals**, which all have **four** sides.

The four interior angles of both a square and a rectangle are all right angles. A consequence of these four interior right angles is: the **opposite sides** of both the squares and the rectangles are **parallel** with each other.

Every square has **four** sides all the **same** length. Every rectangle has **two** opposing **pairs** of sides equal in length.

To calculate the total area of a rectangle or a square, simply multiply the width by the height. For example, a rectangle 3cm wide and 5cm high would have an area of: 3cm × 5cm = 15 square centimeters.

Circle - perfect shape

The distance from the center of a circle to the edge of the circle is the **radius** (r). The diameter, d, of a circle is the full distance across the entire circle through its center.

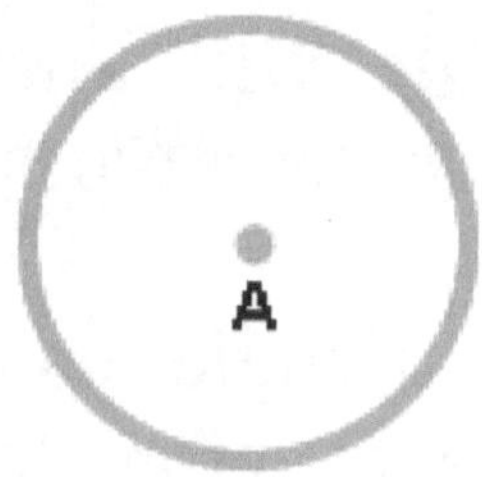

A circle is a shape with all points being the same distance from its center. A circle is named by its center point. Thus, the circle above is referred to as "circle A" as its center is at point A. The **diameter** is **twice** the **radius,** and the circumference is the full distance around the edge of a circle. The number π (pi) is defined as the circumference divided by the diameter. For a circle with radius r: its **circumference** = $2\pi r$. The **area** of a circle can be calculated using the formula: πr^2.

Triangle types

If you draw three dots on a piece of paper and then join them up using straight lines, you have a triangle! Whilst the sum of the four interior angles of a square add up to 360°, the sum of the angles at each vertex of a **triangle** will **always** equal **180°**. There are just three types of triangles:

A **scalene** triangle has three sides all measuring a different length. **All** of the three interior angles are **different** sizes.

An **isosceles** triangle has **two** of the three sides the **same** length. And **two** of the three angles inside the triangle are measured to be the **same**, with the other remaining angle being a different length.

An **equilateral** triangle has **all** three of the triangle's sides measuring the **same** length. All three of the interior angles are exactly **60°**. And the total sum of these three identical interior angles equals 180°.

Trigonometry: triangle study

Trigonometry is the mathematical study of triangles and the relationship between the sides and angles of the triangle. Trigonometry is a mathematics branch which emerged during the 3rd century BC.

In a right-angled triangle the three sides have each been given special names. The side opposite the right angle is called the **hypotenuse**, which is always the **longest** side of the triangle.

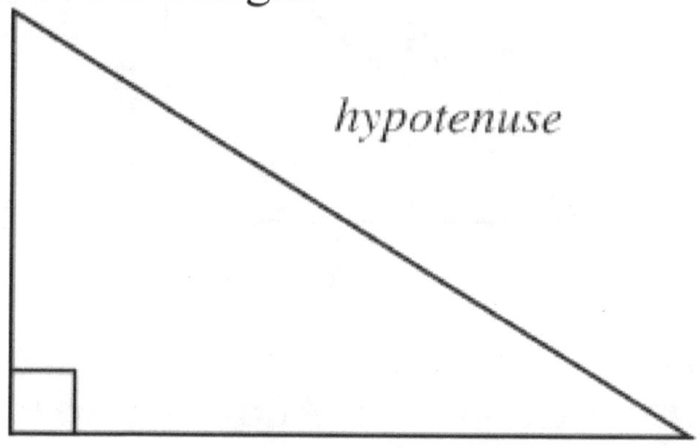

If one of the other angles is known, the side adjacent to this angle is called the **adjacent** side. The side opposite this known angle is called the **opposite** side.

Pythagoras' theorem

Greek mathematician Pythagoras is known for his theorem regarding the lengths of the sides of right-angled triangles. The theorem states if you **square** the lengths of the two shorter sides of a right triangle and add them together, the number you get is equal to the square of the length of the longest side. Best example is the 3-4-5 triangle:

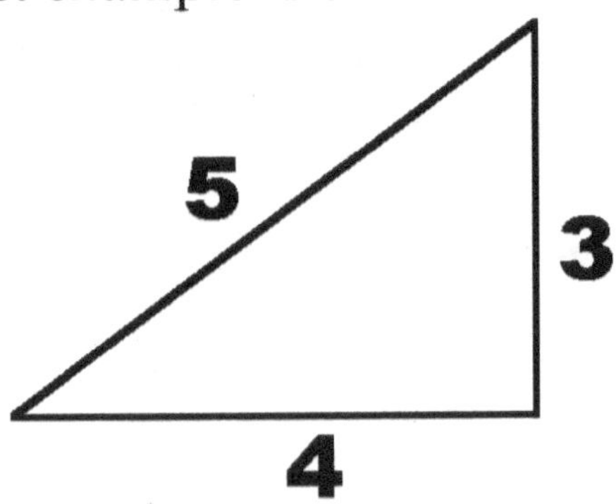

The two **shortest** sides are 3cm and 4cm, so $3^2 = 9$ and $4^2 = 16$. **Add** 9 and 16 to get 25. Pythagoras says this is the length of the **longest side squared**. As 25 is equal to 5^2, we now know the longest side of the triangle measures 5cm.

Pythagoras' theorem

Trigonometry ratios

The three most common trigonometry functions of **sine**, **cosine**, and **tangent** are all simply one side of a triangle divided by another side.

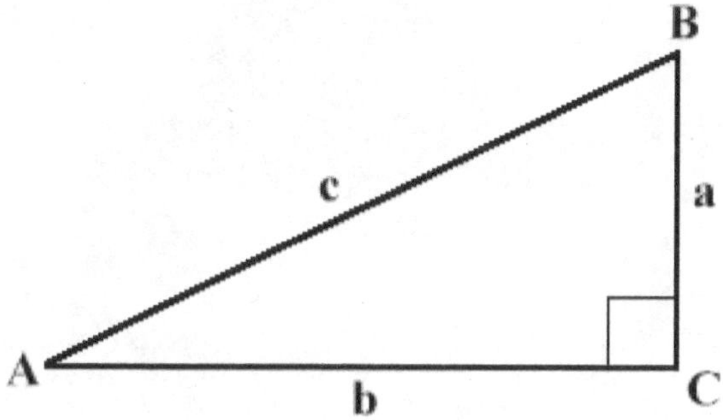

Sine of angle A = measure of opposite side divided by measure of hypotenuse. In the above triangle; **sin** of angle A = (a/c).

Cosine of angle A = measure of adjacent side divided by measure of hypotenuse. In the triangle above; **cos** of angle A = (b/c).

Tangent of angle A = measure of opposite side divided by measure of adjacent side. In the triangle shown above; **tan** of angle A = (a/b).

Pyramid shape

Great Pyramid of Egypt –

one of the ancient wonders of the world

What did the triangle say to the circle?

What's a math teacher's favorite kind of tree?

Geometry

I had an argument with a 90° angle.

It turns out it was right

Parallel lines have so much in common .....

It's a shame they'll never meet

Leonhard Euler – born 1707

Leonhard Euler was a pioneering Swiss mathematician and physicist. He may be the most influential mathematician. Euler gave us **trigonometry**. He worked at the Petersburg Academy and the Berlin Academy of Science. Euler had a phenomenal memory. He once did a mental calculation to settle a dispute between two students whose figuring differed in the fiftieth decimal place!

His mathematical formula, Euler's Formula, is an analysis which establishes the relationship between the trigonometric functions and the complex exponential function. Euler was supreme at **discrete mathematics**, inventing **graph theory** and generating **functions**. He combined his mathematical brilliance with phenomenal concentration. Leonhard Euler is internationally regarded as the best exponent of performing arithmetic (algorism) ever.

Algebra

Algebra is simply a combination of the four elementary arithmetic operations - multiplication, division, addition, and subtraction.

You have gained an understanding of these four arithmetic operations in this book's earlier chapters.

You may like to review the chapters relating to "Add & Subtract" and "Multiply & Divide." Algebra is easier than you think. And algebra is your great mental workout.

A large part of learning algebra is gaining a basic understanding of the relevant graphs. These graphs are not particularly difficult once you grasp exactly what the "*x*-axis" of a graph is, and also what the "*y*-axis" of a graph refers to.

Letters represent variables

Algebra and arithmetic are both branches of pure mathematics. Arithmetic is based entirely on **known** number values. But algebra is based on the concept of **unknown** values. These unknown values, represented by any letters in the alphabet, are known as **variables**. Simply put, algebra is about finding the unknown.

Letters are actually numbers in disguise! These letters are called variables because the numbers they represent can **vary**. This means one or more numbers can be substituted in an algebraic expression.

An algebraic expression is one or more algebraic terms in a phrase. In the phrase, or algebraic expression, of: $3x + 5y$, the letters x and y are both called variables.

Constants are terms in an expression with only numbers. In the expression: $x + y + 8$, the **constant** term is “8”.

Expressions & equations

An algebraic equation is two mathematical expressions joined by the equals sign (=). In algebra, you will learn how to deal with algebraic **expressions,** and you will also learn how to solve algebraic **equations**.

An algebraic expression is anything which involves unknown variables mixed with known numbers. The expressions contain alphabetic symbols and numbers. No matter how complex the expression is it will represent just one number or one quantity. Two examples of an algebraic **expression** are: $2x + 7$; and: $x^2 + 3x + 4$.

It is important to note the expressions do not contain the equals sign. An algebraic equation is like an expression in every way, but the algebraic equation will always include an '**equals**' sign. Two typical examples of an algebraic **equation** are: $2x + 7 = 11$; and: $x^2 + 3x + 4 = 14$.

Keep it balanced

Algebra is very much like a **balance** scale. To keep your scale balanced you will need to do exactly the **same** thing to each side of your algebra equation.

To solve problems in algebra, all you need to do is get all the numbers on one side of the equals sign (=) of your equation. In the example of $y + 4 = 12$, we can either add or subtract numbers from either side of this equation. The important thing here is to make sure you do the **same** thing to each side.

Let's subtract 4 from each side: $y + 4 - 4 = 12 - 4$. Taking away 4 from both sides will simplify things on the left of the equals sign, as +4 −4 means no change! $y = 12 - 4$. As $12 - 4 = 8$, this means $y = 8$.

Always remember: to **keep the balance**, what you do to **one side** of the "=" sign you also do to the "=" sign's **other side**.

Keep it balanced

Graphs can solve equations

A graph is a pictorial representation of an equation. We can define a single point in space relative to another location using both "x" and "y" coordinates. A graph lets us show all of the points whose value of "x" and value of "y" fit some relationship we can represent with an equation.

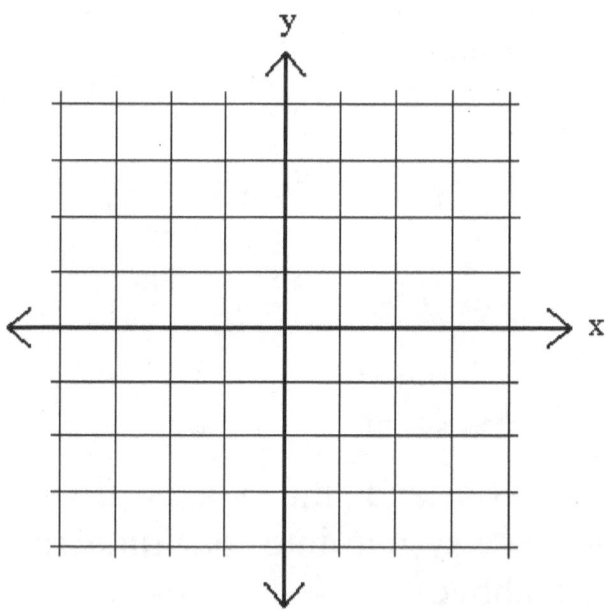

The x-axis and y-axis shown here are the two lines creating the coordinate plane.

X to the left - Y to the sky

The x-axis is a horizontal line. The y-axis is a vertical line. To remember which axis is which, repeat the rhymes: "***x* to the left**," and "***y* to the sky**." Here is the Cartesian coordinate system showing some points plotted.

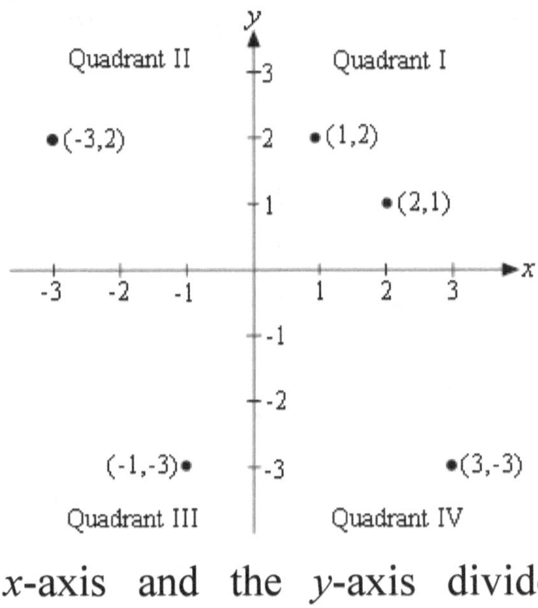

The *x*-axis and the *y*-axis divide the coordinate system into four **quadrants** as shown above.

Quadrant I: $x>0$, $y>0$ (both are positive). Quadrant II: $x<0$, $y>0$. Quadrant III: $x<0$, $y<0$ (negative). Quadrant IV: $x>0$, $y<0$.

Linear equations

If an equation only has terms containing x and nothing else with higher powers, such as x^2 or x^3, then this type of equation is called **linear**. A typical example of a linear equation is:

$$2x + 4y = 8$$

The **letters** "x" and "y" in the above equation are both **variables** whilst the **numbers** "2" and "4" are both **constants**. If you draw a graph showing results of the above equation you will form a straight line, hence the name **linear**.

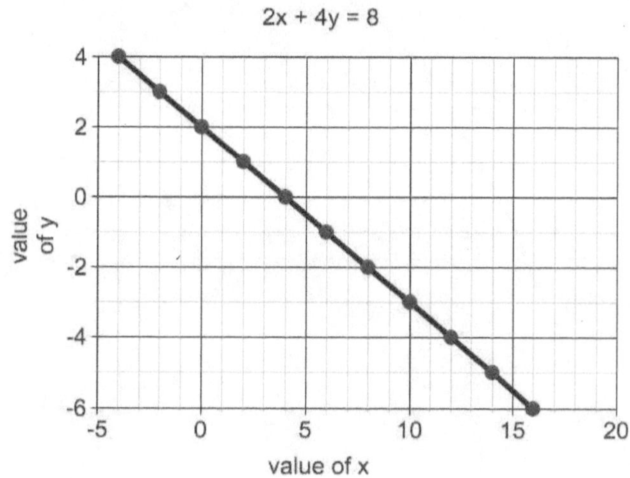

Quadratic equations

An equation whose highest term contains x^2 is called a **quadratic** equation. This means this equation has no terms containing x^3 or anything with higher powers than x^3. One example of a quadratic equation is:

$$x^2 + 2y = 9$$

The **"2"** (in $2y$ above) is a **constant**. If you draw a graph of this quadratic equation, you see a **parabola** shape:

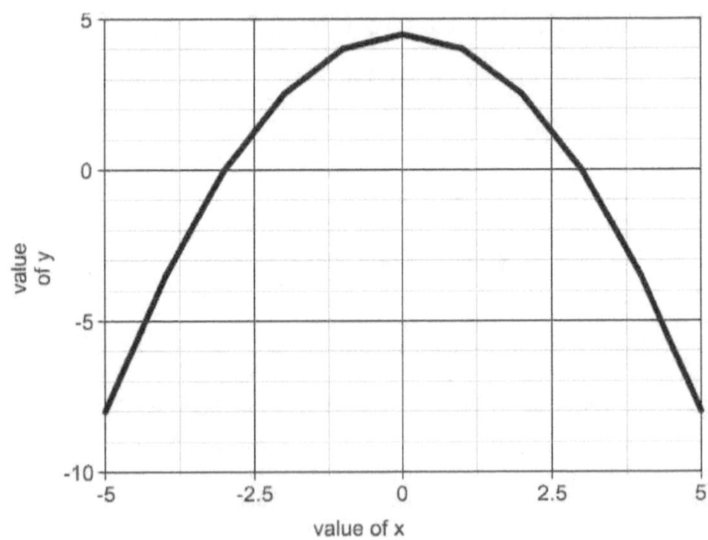

Operations order: PEDMAS

Long ago people devised standard rules when doing calculations. The order in which you do things is important. When there is more than one operation in a calculation, you need to always work in the order of PEDMAS. i.e.

Do things in **P**arentheses (brackets) first

Then do **E**xponents (powers) and roots

Division
Multiplication (these two together

Addition
Subtraction (these two together)

If a calculation involves only division and multiplication, or only addition and subtraction, work from left to right.
PEDMAS stands for the first letters of: **P**arentheses, **E**xponents, **D**ivision and **M**ultiplication, and **A**ddition and **S**ubtraction.

<u>Rearranging equations</u>

It can be useful to **rearrange** an equation to make it much simpler to handle. Before rearranging we need to clarify two things.

Firstly, establish the **order** in which we evaluate an expression. Refer to PEDMAS from the previous page. Secondly, establish which pairs of operations **do** and **undo** something. This means the **operation** and its **inverse** (undo) operation. Inverse operations include; Square ↔ square root, multiply ↔ divide, addition ↔ subtraction.

Two rules for rearranging equations are: <u>Rule 1</u>: You can add, subtract, multiply, and divide by anything, as long as you do the **same thing** to both sides of the "equals" sign. <u>Rule 2</u>: To move or to cancel a quantity or a variable on one side of your equation, you need to perform the "**opposite**" (inverse) operation with it on **both sides** of the equation.

Solve simultaneous equations

Simultaneous equations are a set of equations containing multiple variables. Two linear equations are **simultaneous** if they are considered at the **same time**. An example of $x + 2y = 7$ and $3x + 3y = 15$, are simultaneous equations.

There are just four simple steps to solving the above simultaneous equations:
#1: **Multiply** one equation ($x + 2y = 7$) by a number (3) so one of the unknown terms in the equation becomes the **same** as in the other equation. e.g. $\mathbf{3x} + 6y = 21$.
#2: **Subtract** one equation ($3x + 3y = 15$) from the other ($3x + 6y = 21$) so one of the unknown terms **disappears**. $3x - 3x + 6y - 3y = 21 - 15$. You will note: $3x - 3x = 0$.
#3: **Calculate** the **remaining unknown variable**. $3y = 6$, which means $\mathbf{y = 2}$.
#4: **Use this value** ($y = 2$) in one of the original equations to calculate the other unknown value. You will find $\mathbf{x = 3}$.

Carl Friedrich Gauss – born 1777

German mathematician and scientist Carl Friedrich Gauss is one of our greatest mathematicians of all time. He was born into a poor family but after being a child prodigy, he was awarded a scholarship which assisted him to focus on mathematical research at college.

At age 24 Gauss published *Disquisitiones Arithmeticae*, probably the greatest book of pure mathematics. This 1801 book structured the study of **number theory**. Gauss proved the fundamental theorem of arithmetic, which states every natural number can be represented as the product of primes only one way. He also proved the fundamental theorem of algebra, which shows every integral rational algebraic equation has at least one root or solution. Gauss has often been referred to as the "Prince of Mathematicians." And Gauss himself refers to mathematics as the "Queen of Sciences."

Money Math

American statesman Ben Franklin probably summed it up in the best way when he remarked "Waste neither time nor money, but make the best use of both."

The mathematical principles learnt from this book provide an ideal start to help you become a better manager of your money and make educated financial decisions.

There's always going to be an element of uncertainty regarding most of your financial decisions.

But the basic mathematics you have learned throughout this book should now allow you to understand the nature of this uncertainty. Then you can make the most informed possible choice.

Earn more - spend less

If you are spending more money than you are earning, then you have the recipe for a personal financial disaster. Research shows money problems are actually the number one reason for divorce!

Basic **addition** is used in calculating your total amount of earnings over a full year. To calculate whether or not you have a surplus of money for the whole year, basic **subtraction** is used. From the total amount earned, deduct all the spending during the year from your total earnings.

Optimistically, the annual amount of all money you have earned is higher than the total annual amount you have spent. Earning more than you spend achieves a financial **surplus**, calculated as follows:

Add up your total amount of earnings, then **minus** your total amount of spending to discover your surplus amount.

Three income types

Gross income is the total amount of money you earned **before** any deductions are deducted from the gross amount. If taxes and other deductions are deducted from your pay before you receive your pay, then you will never actually see your gross income.

Net income is the amount of money you have **after** the taxes are withheld and other deductions are deducted from your total pay. Net income is more commonly referred to as **take home pay** as this is the amount which gets credited, or deposited, to your nominated savings account.

Disposable income is the portion of net income left after your obligations have been paid for the current pay period. To determine your disposable income, you must have a listing of all **expenses** payable before your next payday. **Subtract** this total from your net income.

Save more money

Having a **saver attitude** is a very positive financial attitude. It will be beneficial for you mathematically to acquire a habit of not spending all the money you have earned from one or more income sources.

You should think seriously about having **regular savings** stored away for any unexpected situations or emergencies. With enough time to regularly save enough money, you will be more prepared to avoid most situations which may destroy your financial wealth.

Money from your disposable income can either be spent or saved. Your expenses will include both **fixed** expenses and **variable** expenses. Your fixed expenses are those which don't change from month to month such as a mortgage or loan. Your variable expenses might include electricity, water, gas, phone accounts, entertainment, fuel, and groceries.

Adopt an investor attitude

Saving money will not always guarantee your financial success. You may find some larger expenses or unexpected emergencies might eat into the money you have saved. This adversely affects your financial future. Some good savers of money have entered retirement unprepared to fully sustain themselves.

To achieve financial freedom, the attitude you need to adopt is referred to as the **investor attitude**. A person with this type of attitude is primarily a saver of money, but also saves enough of their income to invest in something significant. This something is a valuable investment which will increase a person's income and net worth. Many individuals are considered as great savers. But they can also become great investors if they have the knowledge of a safe and reliable way to convert their savings into a valuable investment which ultimately **increases** their **net worth**.

Time is money

Money and time are interchangeable. You can save one by spending more of the other. If you receive some money today, you are poised to increase the **future value** of this money you have received by investing it and gaining **interest** over a period of time.

Let's examine a straightforward example of mathematically working out the future value of $1000.00 received today: If you invest the $1000.00 at a simple annual rate of 4.5%, the future value of your investment at the end of one year is $1045.00. This amount was calculated by **multiplying** the principal amount of $1000.00 by the interest rate of 4.5%, and then **adding** the $45 interest which was gained to your principal amount:

Future value of investment after first year:
= ($1000.00 × 0.045) + $1000.00
= $1045.00. This will grow each year.

Time is money

Be insured

Everything is exposed to a danger of some type. A person may unexpectedly die, or they could become injured in an accident. Your contents might get stolen, or your home may be destroyed by fire.

The many dangers which result in financial losses are referred to as **risks**. Insurance is concerned with risk. Its main purpose is to reduce the harmful effects of risk, or the fears arising from this risk. Insurance **spreads the risk** over as many people and/or organizations as possible. Therefore, the **premium** you pay is small compared to the possible financial loss.

Insurance companies (insurers) make use of a mathematical principle called the **law of average**, which looks at the average of a large number of cases. Monies paid into an insurance fund are used to compensate policyholders if they suffer a particular misfortune they insured against.

Be insured

Stay out of debt

Staying out of debt is essentially a matter of spending less! Your self-discipline to spend less money combined with the mathematical skills you have gained in this book builds sound financial habits.

The easy credit offered by financiers in today's society tempts many people into spending money they cannot afford. The golden rule for you staying out of debt is **"don't buy** something unless you are **sure you can afford it**."

Create a budget and stick to it. In addition to helping you stay out of debt; the creation of your money budget will sharpen your mathematical skills and ability to calculate a balance. Know what your income is and what your expenses are. From the surplus balance, know what you are able to spend. Pay by cash if you can. Credit cards are handy, but allow you to spend more than you can afford.

Plan shopping

You can practice all of your simple mathematical skills gained from this book each time you shop. We all go shopping for food and essential goods, but some people are able to **save** large amounts on their grocery bills by using their math skills to plan their shopping trips.

Apply your basic addition and subtraction skills every time you walk the aisles in your local supermarket. Two important mathematical concepts for you to use when shopping are: percentages and ratios. If you want to save more money, take time to **compare** the total costs of your favorite meals. **Add** up the cost of all ingredients in a particular meal. Then **estimate** how many servings you may get from this meal, and **divide** by this number. Using simple mathematics, you can compare costs for all of the different foods to determine your best **value** for money in the preparation of your meals.

Double your money!

There is a mathematical formula which allows you to calculate how long it will take to double your money. The formula is called **Rule of 72** and works like this:

To calculate **how long** it will take for your **money to double**, just **divide 72** by the current interest rate. As an example, if the interest rate is 9%, then use the division of: 72 ÷ 9 = 8 years. Thus at the rate of 9% it will take 8 years to double your money. If your interest rate is 6%, just divide 72 by the 6% rate as follows: 72 ÷ 6 = 12 years.

The Rule of 72 can also be used to find the interest rate required to double your money in a specific timeframe. As an example, to double your money in 15 years you use this division calculation: 72 ÷ 15 = 4.8%. Thus an interest rate of 4.8% would be required for you to double your money in 15 years.

Double your money!

Probability

Probability is a mathematical model in which numbers are used to show how likely it is a chance event will happen.

The theory of probability started in the 17th century when Chevalier de Méré, a French gambler, consulted the great mathematician Blaise Pascal about a mathematical way of ensuring he would win at his gambling.

Pascal was able to show the gambler why he was losing, but not how to win! Casinos and bookmakers in the gambling business still use the model first developed by Pascal to make sure they don't go out of business.

We can work out a situation's probability. Assuming the way things can happen are equally likely; probability of an event =

$$\frac{\text{the number of ways it can happen}}{\text{the total number of possibilities}}$$

Odds

Gamblers speak the language of gambling. They talk of ‘odds’ rather than probabilities, but the two are closely related as shown here.

$$\text{Probability of winning} = \frac{\text{number of ways of winning}}{\text{total number of ways of winning or losing}}$$

$$\text{Odds (against winning)} = \frac{\text{number of ways of losing}}{\text{number of ways of winning}}$$

e.g., When rolling a die; the **odds against** obtaining a 2 are 5/1, or ‘5-1’ or ‘5 to 1’. Compare this with the probability, which is recorded as 1/6.

‘Pr’ is used to stand for probability. When rolling a die, the possible outcomes are 1, 2, 3, 4, 5, or 6. The probability of rolling a 2 on the die is one chance in six.

$$Pr\ (2) = 1/6$$

Travel the world!

Travel is one of life's great experiences. Visiting other countries will expand your knowledge of the world and also increase your understanding of many different languages, different currencies, and different cultures. The one constant which remains in every country throughout your travel experience is **mathematics**.

Travel planning is another practical way to use your mathematics in everyday life. Fuel prices are a constant **variable** for all local trips, so it is essential to estimate your fuel costs when planning a long trip.

When travelling to other continents, your mathematical skills will be put to the test as you regularly convert from one **currency** to another. Currency is the form of money you use in each country. There are currently about 180 currencies used around the world, and their respective **exchange rates** fluctuate daily.

Travel the world!

Frankfurt Airport

Congratulations!

You made it! Your journey through this publication has reached its climax with this last vital chapter regarding your money handling. You have successfully completed this Best Practice challenge!

This Best Practice book is designed to be your ideal introduction to the rewarding subject of Mathematics. But your journey has really only just begun!

So don't stop your mathematical journey here. You can now dig a lot deeper into all of the varied and interesting branches of mathematics. And you can also add to your existing knowledge of the essential basics of mathematics.

Then you can challenge yourself to continue to learn even more and to develop your valuable mathematics skills.

You will feel your self-confidence really grow rapidly as your mathematical competency grows. You can count on it!

Math in the 21st century

Today, mathematics plays a leading role in science, education, and industry. Applied Mathematics provides innovative solutions to many of our needs for the 21st century. The application of mathematics to the other sciences leads us to great and exciting new mathematical discoveries.

Applied Mathematics looks for important connections with other disciplines which may inspire useful and interesting mathematics. It is a branch of mathematics which concerns itself with mathematical techniques used in the application of mathematical knowledge to a number of other domains or sciences.

Innovative mathematical reasoning may lead to many exciting new insights and applications in these following areas to name just a few: **Climate Change, Physics, Biology, Finance, Statistics, Information Science, Market Research, Economics, Engineering, Banking,** and **Medical Research.**

Alan Turing – born 1912

English mathematician and philosopher Alan Turing was highly influential in the development of theoretical computer science, providing a formalization of the concepts of algorithm and computation with the Turing machine, which can be considered a model of a general-purpose computer. He is widely considered to be the father of theoretical computer science and artificial intelligence.

In 1938, Alan obtained his PhD from Princeton University. During World War II, Turing worked for the Government Code and Cypher School at Bletchley Park, the code-breaking centre which produced Ultra intelligence. For a time, he led Hut 8, the section responsible for German naval cryptanalysis. Here, he devised a number of techniques for speeding the breaking of German ciphers. Alan Turing played a crucial role in cracking intercepted coded messages that enabled the Allies to defeat the Axis powers in many crucial engagements, including the Battle of the Atlantic.

Women in mathematics

All of the great mathematicians listed earlier in this book are men. But math is all about a journey of discovery, not someone's gender. Student should never feel they cannot do something because of their gender.

Stories of women in mathematics are countless but mostly obscured. There are many women in history who managed to make a name for themselves, despite gender discrimination. But even those women have not been widely celebrated.

Inequality with access to mathematical education is not only gender based but it is also dependent on factors such as geography, class and cultural perceptions.

The following three women have helped to establish a path to inclusive global mathematics. More female role models are now gaining more visibility, and this will lead to inspiring a new generation of women mathematicians.

Sophie Germain – born 1776

Sophie Germain was a French mathematician and physicist. Despite opposition from society and her parents (due to her gender), Sophie persevered and went to greatly contribute to mathematics. Due to unfair gender restraints at the time, she was not able to pursue a career in math. However, this did not stop Sophie! She worked tirelessly on mathematics independently throughout her life.

Through Germain's hard work, she advanced the field of number theory. Sophie was also one of the pioneers of elasticity theory, becoming the first woman to win the grand prize from the Paris Academy of Sciences for her essay on the subject. Sophie proved to people that females are incredible mathematicians. Although not allowed to attend university, she became the first woman to make important contributions to mathematical research. Sophie Germain carved out the way for women today.

Katherine Johnson – born 1918

Katherine Johnson was born in USA. As a mathematician, she calculated and analysed the flight paths of spacecraft during the US space program. Her work helped send astronauts to the Moon. In 1939, Johnson was selected to be one of the first three African American students to enrol in a graduate program at West Virginia University. Later she was part of a group of NASA employees called "computers." This was a group of African American women who excelled in mathematics and problem-solving.

In 1962, as NASA prepared for John Glenn's orbital mission, Katherine was asked to do the work that she became famous for. The flight's complexity required construction of a huge communications network. NASA's astronauts were wary of putting their lives in the care of electronic calculating machines. Glenn asked engineers to "get the girl" - Johnson – to run the numbers by hand on her mechanical machine. Glenn stated, "If she says they're good, then I'm ready to go." The flight was a success.

Shakuntala Devi – born 1929

Shakuntala Devi was an Indian writer and mental calculator. She was popularly known as the "Human Computer." Devi was a precocious child, and she demonstrated her arithmetic skill at the University of Mysore without any formal education. Here, Shakuntala strived to simplify numerical calculations for students. Devi wrote many books including texts about mathematics and puzzles. In 1977, she computed the 23rd root of a 201-digit number in 50 seconds. It took a computer 62 seconds to find the answer.

Devi travelled to several countries around the world, demonstrating her arithmetic talents. She explained many of the methods she used to do mental calculations in her 1977 book *Figuring: The Joy of Numbers*. Skakuntala said, "At three, I fell in love with numbers. It was sheer ecstasy for me to do sums and get the right answers. Numbers were toys with which I could play." Devi became rich and famous using her talents. She achieved something people from even the most privileged backgrounds couldn't achieve.

Nature of Mathematics

Here are some inspiring and thought-provoking quotes about mathematics.

"Mathematics, rightly viewed, not only possesses not only truth but supreme beauty."

Bertrand Russell

"Pure mathematics is, in its way, the poetry of logical ideas."

Albert Einstein

"Mathematics is the most beautiful and most powerful creation of the human spirit."

Stefan Banach

"Mathematics is a language."

Marcus du Sautoy

"Mathematics compares the most diverse phenomena and discovers the secret analogies that unite them."

Joseph Fourier

Motivation

Two motivating quotes and two funny quotes.

"If I were beginning my studies, I would follow the advice of Plato and start with mathematics."

Galileo Galilei

"The definition of a good mathematical problem is the mathematics it generates rather than the problem itself."

Andrew Wiles

"If there is a 50-50 chance that something can go wrong, then nine times out of ten, it will go wrong."

Paul Harvey

"Arithmetic is numbers you squeeze from your head to your hand to your pencil to your paper till you get the answer."

Carl Sandburg

Books in series:

Be Better

Live better, plan better, learn better, look better, love better, eat better, work better, and much more! This is your personal **growth** bible.

Planets tour

Take an amazing journey to all planets of our Solar System. Discover **highlights** of each planet and the **unique** qualities of each planet.

Universe tour

Take the **ultimate journey** through our vast universe. Visit planets, moons, stars, and galaxies. Travel to the most distant galaxy.

Bible highlights

Take a **tour** and follow the Bible with selected verses. Learn **valuable** lessons from 100 verses, including the world's favourite Bible verse.

French tongue

French is presented in an easy-to-follow format. Learn essential words and phrases. And learn the correct pronunciation of words and phrases.

Born in 20th century

All quotes are from people born in the 20th century. More than a thousand profound events are listed. One hundred years in your hands!

Books in series (2)

Acts, Romans, Corinthians

Acts, Romans & Corinthians are the sequel to the Gospels. Luke and Paul wrote these books because Christians wanted to know Jesus more.

Confucius says

The profound quotes of Confucius affect one quarter of humanity and still echo in the mind. All these sayings are universal and timeless.

Italian tongue

The passionate language of Italian is presented for you in an easy-to-follow format. Prepare for a trip to Italy – the holiday of a lifetime.

Jesus says

Jesus is the most influential person in all of human history. Adopt these best practices in your life, as taught by the son of God, Jesus.

New Testament nuggets

Read the greatest collection of spiritual verses written in the 1st century. All 27 New Testament books are represented with profound verses.

Psalms & Proverbs

Psalms and Proverbs are the "core" of the Bible. Profound verses give valuable life lessons. Read more of the Old - a Testament full of gold!

Books in series (3)

New You

Want to build a new you? Develop a new plan, gain new energy, and build a new mind, body, and spirit. Create and nurture a new you.

Tennis essentials

Practice ball sense and footwork to build tennis skills for a lifetime of **fun** and **fitness**. Learn the **essentials**. Practice the **shots**. Play the **game**!

OLD into NEW

The New Testament is our divinely inspired commentary on the Old. It's beneficial for us to see how authors of the New understood the Old.

Echoes of wisdom

This is a fascinating compilation of inspired quotes from the greats of our modern history. A treasure trove of timeless quotes from our past!

Spanish tongue

Spanish is the main language in 21 countries. Learn essential words and phrases. Practice the correct pronunciation of words and phrases.

She said

Over two hundred women share inspirational quotes with you. Profound quotes come from Cleopatra to the Queen, to Greta Thunberg.

Books in series (4)

German tongue

German is the **most spoken** language within the European Union. Learn essential words and phrases and use their correct pronunciation.

Winning

What are the qualities of a winner? How do you experience winning more often? Success is an issue of character, not performance. Get *Winning* before your opponent reads it!

"I am"

Affirmations inspire and motivate you. You are encouraged to pick "*I am*" positive affirmations. They build a winner's mentality and qualities.

Greek greats

Philosophy is a Greek creation. Ancient Greeks used it to make sense of the world. Learn from profound quotes of 25 leading thinkers.

Five words or six words

Short quotes are packed with wisdom. Quotes help you reveal and assess your assumptions, your values, and your beliefs. Five words to motivate you. Six words to change your life.

www.ingramcontent.com/pod-product-compliance
Lightning Source LLC
LaVergne TN
LVHW090409210926
848480LV00002B/513